MANAGING URBAN FUTURES

Managing Urban Futures

Sustainability and Urban Growth in Developing Countries

Edited by

MARCO KEINER, MARTINA KOLL-SCHRETZENMAYR,
WILLY A. SCHMID
Swiss Federal Institute of Technology (ETH),
Zurich, Switzerland

ASHGATE

Published by
Ashgate Publishing Limited
Gower House
Croft Road
Aldershot
Hampshire GU11 3HR
England

Ashgate Publishing Company
Suite 420
101 Cherry Street
Burlington, VT 05401-4405
USA

Ashgate website: http://www.ashgate.com

British Library Cataloguing in Publication Data
Managing urban futures : sustainability and urban growth in
    developing countries
    1.City planning - Developing countries 2.Urbanization -
    Developing countries 3.Cities and towns - Developing
    countries - Growth 4.Sustainable development - Developing
    countries
    I.Keiner, Marco, 1963- II.Koll-Schretzenmayr, Martina,
    1967- III.Schmid, Willy A.
    307.1'416'091724

Library of Congress Cataloging-in-Publication Data
Managing urban futures : sustainability and urban growth in developing countries /
edited by Marco Keiner, Martina Koll-Schretzenmayr, and Willy A. Schmid.
    p. cm.
    Includes bibliographical references and index.
    ISBN 0-7546-4417-0
    1. City planning--Developing countries. 2. Urbanization--Developing countries. 3.
Urban ecology--Developing countries. I. Keiner, Marco, 1963- II. Koll-Schretzenmayr,
Martina, 1967- III. Schmid, Willy A.

    HT169.5.M26 2005
    307.1'216'091724--dc22

                                                            2005003240

ISBN 0 7546 4417 0

Printed in Great Britain by MPG Books, Bodmin, Cornwall.

Contents

v

List of Figures

List of Tables

List of Contributors

Caldeira, Teresa, Prof. Dr., University of California, Anthropology, 5275 Social Science Plaza, Irvine, CA 92679-5100, USA, tpcaldei@uci.edu

Feiner, Jacques P., Dr., Institute for Spatial and Landscape Planning, Swiss Federal Institute of Technology ETH, Hönggerberg, CH-8093 Zurich, Switzerland, feiner@nsl.ethz.ch

Friedmann, John, Prof. em. Dr., University of British Columbia, School of Community and Regional Planning, 1933 West Mall, 2nd Floor, Vancouver, B.C. V6T 1Z2, Canada, jrpf@interchange.ubc.ca

Gilbert, Alan, Prof., University College London, 26 Bedford Way, London WC1H 0AP, United Kingdom, a.gilbert@geog.ucl.ac.uk

Ginkel, Hans J.A. van, Prof. Dr., United Nations University Centre, 53-70 Jingumae 5-chome, Shibuya-ku, Tokyo 150-8925, Japan, Rector@hq.unu.edu

Holston, James, Prof. Dr., University of California, Anthropology, 9500 Gilman Drive, La Jolla, CA 92093-0532, USA, jholston@ucsd.edu

Keiner, Marco, Dr., Institute for Spatial and Landscape Planning, Swiss Federal Institute of Technology ETH, Hönggerberg, CH-8093 Zurich, Switzerland, keiner@nsl.ethz.ch

Koll-Schretzenmayr, Martina, Dr., City and Landscape Network, Swiss Federal Institute of Technology ETH, Hönggerberg, CH-8093 Zurich, Switzerland, schretzenmayr@nsl.ethz.ch

Kunzmann, Klaus R., Prof. Dr., University of Dortmund, Department of Spatial Planning, PO Box 500500, D-44221 Dortmund, klaus.kunzmann@udo.edu

Leaf, Michael, Prof. Dr., University of British Columbia, Center for Human Settlements, 1933 West Mall, 2nd Floor, Vancouver, BC V6T 1Z2, Canada, leaf@interchange.ubc.ca

Marcotullio, Peter J., Dr., United Nations University, Institute of Advanced Studies (UNU/IAS), International Organizations Center, 6F, Pacifico-Yokohama, 1-1-1 Minato Mirai, Nishi-ku, Yokohama 220-0012, Japan, pjmarco@ias.unu.edu

McGranahan, Gordon, Dr., Director of Human Settlements Programme, International Institute for Environment and Development, 3 Endsleigh Street, London WC1H 0DD, United Kingdom, Gordon.McGranahan@iied.org

Perlman, Janice E., Prof. Dr., Founder and President, The Mega-Cities Project, Inc.; Visiting Scholar, Urban Planning Program, Graduate School of Architecture, Planning and Preservation, Columbia University, 1172 Amsterdam Avenue, New York City, NY 10027, USA, jp2328@columbia.edu and Jperlman@worldbank.org

Rakodi, Carole, Prof. Dr., University of Birmingham, International Development Department, School of Public Policy, University of Birmingham, Edgbaston B15 2TT, United Kingdom, c.rakodi@bham.ac.uk

Salmerón, Diego, Institute for Spatial and Landscape Planning, Swiss Federal Institute of Technology ETH, Hönggerberg, CH-8093 Zurich, Switzerland, salmeron@nsl.ethz.ch

Sandercock, Leonie, Prof. Dr., University of British Columbia, School of Community and Regional Planning, 33-6333 Memorial Road, Vancouver, BC V6T 1Z2, Canada, leonies@interchange.ubc.ca

Sassen, Saskia, Prof. Dr., University of Chicago, Ralph Lewis Professor of Sociology, The University of Chicago Law School, 1111 E. 60th St., Chicago, IL 60637, USA, s-sassen@uchicago.edu

Schmid, Willy A., Prof. Dr., Institute for Spatial and Landscape Planning, Swiss Federal Institute of Technology ETH, Hönggerberg, CH-8093 Zurich, Switzerland, schmid@nsl.ethz.ch

Foreword – Managing the Future City

Gordon McGranahan

Cities have always displayed the heights of our ingenuity, and the depths of our depravity. They are social in the extreme, and yet induce individualism and anomie. They are motors of economic growth, and breed terrible inefficiencies. They are the crucibles of citizenship, and the terrains of class conflict. They are sites of spontaneity that are forever being planned and re-planned. They increase local diversity, even as they drive globalization. They embody environmentalists' worst nightmares, and their greatest hopes for the future. And now that half of the world's population is urban, the distinction between urban and rural is itself eroding. Such apparent contradictions are the stuff of urban development and management.

The notion that cities are managed by urban planners with master plans has always been a fiction, and is now only rarely acted out in real cities. The urban managers we encounter in this book range from *favela* dwellers of Rio de Janeiro to the wildly ambitious designers of Brasília; from local politicians in Mombasa to the international financiers of London, New York and Tokyo; from social movements in Latin America to networks of Internet users in China. Few of these urban managers wait for their marching orders – for better or worse they get on with it, plan or no plan, usually creating futures quite different from what was intended.

By and large, the cities in low-income countries face the greater difficulties, and that is where the book focuses. Unfortunately, affluent cities provide few answers that are relevant to their less well-off cousins. To the extent that they have shifted their burdens over time and space, affluent cities need to be challenged rather than copied. In any case, one of the few generalizations one can make about poor cities is that they cannot afford to be managed like affluent cities. But perhaps most important, there is little point in searching for an ideal model of how cities ought to be managed. What is far more important is to develop and debate better understandings of how cities develop and change, and that is what this book seeks to contribute to. It is on the basis of such understandings that the future cities can, hopefully, become better managed.

The authors are almost all long time urbanists, who provide a wide range of viewpoints on cities, despite the fact that they are all working out of Northern research institutions. They focus on different regions (Latin America, Africa, Asia), different challenges (the environment, globalization, multiculturalism), and different scales (local residents, cities, networks of cities). Yet they are united in a common belief that lessons from years of work studying cities can be distilled into chapters that are understandable to the educated reader.

The result is a book with the character of a vibrant and successful city, full of different panoramas and (in-)sights, yet forming a coherent whole. Some will prefer the more formal and safer chapters, others the more exotic. But every chapter has something to offer the urban managers of today, and perhaps more important tomorrow.

Acknowledgements

We wish to thank the people and organizations who helped make this book possible. First of all, we owe special thanks to the contributors, without whose sustained commitment the volume would not have been possible. It has been our special privilege to work with them. We want to extend a note of special appreciation to our institution, the *Netzwerk Stadt und Landschaft* at the Swiss Federal Institute of Technology (ETH), Zurich, for providing the resources for this intellectual pursuit. In addition, we would like to thank Prof. Hans J.A. van Ginkel, Rector of the United Nations University (UNU) and Dr. Thomas Wagner, Zürich, for acting as co-organizers of the Conference on 'Sustainability and Urban Growth in Developing Countries' held 31 October to 3 November 2004 at Monte Verità (Ascona, Switzerland) that inspired this book. Last but not least, we would like to thank all those who attended the conference at Monte Verità and made it fruitful thanks to their contribution and discussion.

On a more individual level, we are especially grateful to Arley Kim and Beverly Zumbühl for providing indispensable support in seeing the book through the editing process and suggesting helpful improvements for the original text. We owe a considerable debt to everyone who helped in the production of the book especially Valerie Rose and Sarah Horsley of Ashgate Publishing. Special thanks to Anita Schürch, Editorial Assistant, who gave invaluable help, support and form to the undertaking, and to Andreas Gähwiler and Oswald Roth, Graphic Designers in the Reprographic Unit of the *Netzwerk Stadt und Landschaft* at ETH Zurich.

Finally, we would like to thank the United Nations University Press for permission to include Figures 1.2 and 1.3.

Every effort has been made to contact copyright holders for their permission to reprint material in this book. The publisher would be grateful to hear from any copyright holder who is not acknowledged here and will rectify any errors or omissions in future editions of this book.

Marco Keiner
Martina Koll-Schretzenmayr
Willy A. Schmid

Introduction

Marco Keiner, Martina Koll-Schretzenmayr and Willy A. Schmid

UNDP (1994) defines human development as 'a process of enlarging the choices for all people in society'. More specifically, sustainable human development '...places people at the centre of the development process and makes the central purpose of development as creating an enabling environment in which all people can enjoy a long, healthy and creative life'. Thus, it seeks to ensure that everyone, particularly the poor and the vulnerable, benefit from economic growth through empowerment by expanding people's capabilities and choices without compromising those of future generations; encouraging better ways in which communities and individuals cooperate and interact; and promoting equity in terms of income, capabilities, opportunities, and the security of livelihood.

The seven Principles of the Sustainable City (CIFAL 2004) present the dichotomy of local decisions and planning actions and their relation to meeting basic environmental, social and human needs, or 'the vision of the whole', a principle that lies at the core of urban existence. For example, local resource use should respect nature and the environment at large, local self-sufficiency is undeniably tied with understanding the links and needs of the world over, and employment is one aspect of building citizenship. The urban reality of living together requires thinking in larger terms that ultimately, by definition, acknowledge sustainability.

The future is unquestionably urban, that is, by 2020 more than half of the globe's eight billion people will be living in cities. There will be 27 mega-cities and more than 500 additional cities will have more than one million inhabitants. However, the urban present is just the beginning of the urban future. By using satellite images and global urban indicators, UN-HABITAT's Global Urban Observatory is monitoring and evaluating urban conditions and trends. Their findings are quite depressing:

... Poorly managed cities and towns contribute to unsustainable production and consumption patterns. They also generate unmanageable wastes, which negatively impact on land and water resources as well as on the atmosphere. Sustaining healthy environments in the urbanized world of the 21st century represents a major challenge for human settlements, development and management. Social gains have lagged behind economic gains. A large proportion of the world's population remains deprived of basic services such as water supply and sanitation. The marginalized population is particularly large in those countries of the South where economic progress was slow ... [resulting] in the proliferation of slums and squatter settlements. ... Most cities in both developed and developing regions experience a polarization of their populations into affluent and poor neighbourhoods. Trends are often towards further segregation rather than social integration of the poor and marginalized urban population groups. Poverty is a growing problem worldwide both in the rural and urban areas and impacts negatively on human settlements. At least 600 million urban dwellers in Africa, Asia and Latin America live in 'life-and-health threatening homes' and neighbourhoods because of the very poor housing and living conditions and the lack of adequate provision for sanitation, drainage, removal of garbage and health care. An increasing number of the urban poor are also homeless in both developed and developing countries, with current estimates of the homeless population being over 100 million. In many countries, the rates of urbanization exceed the capacity of national and local governments to plan and organize this transformation. As a result, new forms of urban poverty have emerged, manifested through poor housing conditions, insecure land tenure, urban crime and homelessness. Moreover, poorly managed cities have negative impacts on environmental conditions. In most countries in Africa, Asia and Latin America, the absolute level of resources available to local governments is seldom adequate to provide even the most minimal level of services.[1]

By going through the definition of 'sustainable human development' and the seven Principles of the Sustainable City, remembering all the effort done by international organizations, development aid institutions, NGOs and the World Bank to combat the problems of the human condition, and acknowledging the dismal findings of the Global Urban Observatory leaves us with the question, 'If the sustainable city was the utopia of the 20th century, what do we do now in the new Millennium?'. Can the obvious gap between the unsustainable development of cities and the normative postulations of how sustainable cities should exhibit be bridged?

The aim of this book is to present a closer look at the management of urban future. By analyzing the Asian, African and Latin American urban challenges and reflecting on processes and driving forces that set new scales of space and time influencing global urban development, the framework is provided to present socio-political analysis and case studies that discuss the urban present and prospect of cities and city-regions related to the transformations of state, space and society in Brazil and China as well as the future relationship between North and South, between multiple urban cultures, and urban growth and 'evolutionable' cities.

Presentation of the Book

The book is divided in four parts. Part I, entitled *The Challenges of Urban Development Processes* gives an overview on the ongoing urban transition in Asia, Latin America and Africa. Part II, *New Scales in Space and Time* highlights the global interconnectedness of urban environmental development over space and time. Part III, *Transforming State, Space and Society: Political Reform, Promoting Citizens Rights and Poverty Alleviation* explores which models of governance to follow and how changes in policy and citizen's participation influence the outcome of planning, followed by two different case studies on urban poor and on the application of planning alternatives. The final Part IV, *Coming Together*, emphasizes on the inclusion of citizens and the concept of 'evolutionable' cities and treats the question whether solutions from the North can be applied in the South.

Part I: The Challenges of Urban Development Processes

The first chapter on *Asian Urbanization and Local and Global Environmental Challenges* by Hans J.A. van Ginkel and Peter J. Marcotullio explores current economic and urbanization transformations in Asian cities. Main local and global environmental challenges that have accompanied these changes are focused on. The 'infrastructure time-bomb', defined as the overburdening of old, out-of-date, and overused basic infrastructure, is examined in the Asian context.

A hypothesis of the chapter by Alan Gilbert on *Sustaining Urban Development in Latin America in an Unpredictable World* is that future urbanization will only increase the burden on these structures until they collapse if this situation is not addressed. Although the metropolitan centers in Latin America are now growing relatively slowly, at least in comparison with the days of rapid cityward migration in the 1950s and 1960s, they face a major challenge in terms of overcoming poverty, generating employment, providing adequate infrastructure and services, and solving the problems associated with growing social inequality. Urban governments have to address the problems associated with poverty, but they also have to solve the same kinds of problems that face all modernizing cities – the diseases of traffic congestion, pollution, inequality and crime. Moreover, the external economic context in which Latin American cities must develop is not a good one. Today, national economic development in many countries of Latin America largely depends on decisions taken in the world's financial centers, that is, London, New York and Tokyo. Thus, the New Economic Model, meant to improve Latin America's international competitiveness, has not entirely shown the expected results. Latin America's large cities are in urgent need for affordable housing for migrants from the rural areas, improvements in public transport infrastructures, and they also have to deal with high levels of crime. To sum up, the challenges facing the cities of Latin America are formidable due to widespread poverty and uncontrolled urban expansion and general lack of self-sufficiency in an increasingly globalized world.

Sub-Saharan African towns and cities, too, are often portrayed as being in crisis. The challenges that they face include rapid population growth unaccompanied by industrialization or economic growth, lack of economic dynamism, governance failures, severe infrastructure and service deficiencies, inadequate land administration, poverty, and social breakdown. However, urban centers continue to grow and function despite the severity of these challenges, a phenomenon requiring a closer look. Carole Rakodi's chapter on *The Urban Challenge in Africa* provides a broad overview of approaches to administering towns and cities and the alternatives that have developed in some cases, in order to identify pointers to more effective governance in the future. It concludes that African urban centers are indeed facing serious challenges and some of them are in crisis, especially those affected by large-scale civil conflict and government collapse.

Part II: New Scales in Space and Time

Globalization, in economic terms, does not lead to an even distribution of economic power and employment worldwide. The effects of economic globalization lead to the concentration of investment, financial power and business headquarters in a few big 'global cities' (Sassen 2000b). This concentration is due to increased and improved information technologies and hypermobile fluxes of capital at a global scale. The information society's dream of the 'global village', in which the location of the production of information is independent from space and international enterprises become virtual, is not fulfilled yet. In contrary, as informalization is leading to the production and distribution of goods and services at a lower cost, the multinationals – even the most advanced information industries – are more than ever localized in large cities, and the inequality in the concentration of strategic resources and activities between major international financial and business centers and other cities in the same country has sharpened. In addition, the gap between the economic centers and their hinterlands continues to widen because global cities interact and network among themselves. This spatial trend of economic activities worldwide has contributed to new forms of territorial centralization. In her chapter *The Global City: Strategic Site, New Frontier*, Saskia Sassen explores this new geography of centrality and marginality engendered by the processes of globalization.

The chapter *A Question of Boundaries: Planning and Asian Urban Transitions* by Michael Leaf addresses problems associated with the conventional boundaries used in analyzing processes of urbanization and devising means for planning and policy responses. With reference to examples from China, Indonesia, and elsewhere among the rapidly urbanizing countries of Pacific Asia, 'the boundary question' is examined with regard to three interlinked aspects of urbanization and urban change: societal urban transition, resulting patterns of spatial transformation, and the role of urban planning.

Cities in the Asia-Pacific region have undergone rapid and extensive transformations over the past few decades. These changes are associated with unique and sometimes extreme environmental conditions. The chapter *Shifting Drivers of Change, Time-space Telescoping and Urban Environmental Transitions in the Asia-Pacific Region* presented by Peter J. Marcotullio argues that the observed trends are related to time and space effects of new drivers of environmental change. Time- and space-related impacts have transformed the timing, speed and sequencing of environmental transitions such that challenges are appearing sooner, growing faster and emerging more often simultaneously than those previously experienced by industrialized cities. The term 'time-space telescoping' is used to identify these processes. The chapter by Peter J. Marcotullio outlines the theory of 'time-space telescoping', identifies the changes in some of the major drivers, and examines some of their impacts.

Part III: Transforming State, Space and Society: Political Reform, Promoting Citizens Rights and Poverty Alleviation

The transfer of power and decision-making to the civil society at large appears to be of utmost importance for establishing effective democratic governments. John Friedmann examines its historical underpinnings as well as theoretical and practical implications in *Civil Society Revisited: Travels in Latin America and China*, pointing out that the understanding of this term may vary considerably from one country to another. The chapter visits a number of Latin American countries – Brazil, Chile, and Peru – where civil society first emerged in the 1960s and 1970s during military dictatorships, the turn to a neo-liberal economic policy, and catastrophic levels of unemployment. John Friedmann highlights four established models of civil society, the Tocquevillean model of associative democracy, the Habermasian model of public sphere, the Castellsian model of social movements, and finally the Gramscian model of counter-hegemonic practices. He then poses the question if any of the four models he has identified can be applied to China and with what consequences.

Not only former socialist states, but also countries like Brazil underwent a period of strong and centralized state intervention in the production of urban space. In the last half century, the Brazilian state consolidated and then destroyed a modernist model for the production of urban space. According to this model, best crystallized in the construction of Brasília, the state produces urban space through centralized master plans that are conceived as instruments of social change and economic development. The role of government is both to articulate these plans and create the means for their realization. As Teresa Caldeira and James Holston outline in their contribution *State and Urban Space in Brazil: From Modernist Planning to Democratic Interventions*, the total top-down planning approach that imposes urban 'solutions' is increasingly contested by civil society, claiming plans that are based on democratic citizenship and reducing the role of government to that of a manager of localized and private interests in the development of the cityscape. The new model considers that plans should both be based on and foster the exercise of citizenship and, thus, is an explicit expression of the democratization process that has been transforming Brazilian society since the 1970s. The authors contrast these two models of total planning by the state with the people-centered management of urban space and reveal that more democracy in a neo-liberal environment does not necessarily lead to the decrease of social segregation.

The chapter *The Chronic Poor in Rio de Janeiro: What has Changed in 30 Years?* by Janice E. Perlman reports on findings about chronic poverty, persistent inequality, and exclusion and citizenship based on panel studies over time in three *favelas* (squatter communities) in Rio de Janeiro. The author of *The Myth of Marginality: Urban Politics and Poverty in Rio de Janeiro* originally interviewed 750 *favela* residents in 1968–69, collecting qualitative and quantitative data and, since 1999, has been conducting a re-study of the same people and communities. Comparing within and across generations, the author shows that, although there have been notable improvements in the consumption of collective urban services, household goods, and years of schooling over the past three decades, is the city suffers from even greater unemployment and inequality than before. The stigma of *favela* life is reflected in a lack of return on educational investment and extraordinary earning differentials between *favelados* and non-*favelados* in the same areas of the city. Among the barriers to livelihoods is the expropriation of the *favelas* by drug dealers in complicity with police, politicians, and international criminal networks. The violence between rival gangs and police creates a pervasive sense of fear and dampens the once vibrant social life and social capital of the communities. Despite the promise of a greater 'voice' and sense of citizenship with the end of the dictatorship in 1984, *favela* residents feel they have less bargaining power now and are more severely excluded from job opportunities.

China, the emerging economic power, is full of contrasts. On the one hand, it is characterized by fast-growing cities with skyscrapers, but on the other hand, its rural areas seem to be very far from benefiting from the current development boom. This contradiction is focused on in the chapter *Unsustainable Trends in Spatial Development in China: Situation Analysis and Exploration of Alternative Development Paths Demonstrated by Case Studies of Kunming (Urban) and Shaxi Valley (Rural)* by Jacques P. Feiner and Diego Salmerón. The authors highlight the need for a comprehensive planning approach in large city regions and parallel efforts to improve economic and settlement conditions by preserving cultural heritage sites in the rural milieu.

Part IV: Coming Together

The chapter of Leonie Sandercock, *Sustaining Cosmopolis: Managing Multicultural Cities*, looks at a range of possible policy responses addressing the integration of immigrants, drawing primarily from examples of cities in Canada and Australia, countries that have officially espoused multiculturalism since the 1970s. Policy research in recent years has made clear the economic and demographic necessity of developed countries on an increasing flow of immigrants. Countries that have hitherto not thought of themselves as countries of immigration are becoming more and more so, thereby turning, empirically speaking, multicultural and multiethnic. How can this process be managed in ways that achieve peaceful coexistence and the forging of new, intercultural forms of urban life, rather than the repression and denial of rights to immigrants, resulting in increasing social tensions? Seven policy directions are considered: from political and policy support systems to the culture and practices of municipal workers; from reform and innovation in the realm of social policy to a better understanding of how urban policies can address and accommodate cultural differences; and from new models of citizenship to a recognition of and preparedness to work with the emotions that drive conflicts over integration. If the 'mongrel cities' (Sandercock 2003) of the 21st century are to be socially sustainable, their citizens, city governments, and city-building professions need to work collaboratively on all these fronts.

Urban growth in developing countries is a challenge to the management capabilities of national and local authorities. Problems of steering urban development occur not only in the so-called 'mega-cities' of eight to ten million inhabitants or more but also in medium-sized cities in developing countries. These cities can be characterized by rapid population growth, an economy dependent on the informal sector, rampant poverty, widespread informal housing, and basic environment, public health, and governance problems. All of these cities consume more and more resources and overstep local, regional, and global carrying capacities. It is clear that the growth of population and urban areas cannot continue. Sustainable development is an answer, but the implementation of this concept into policy often occurs too late. Urban growth in developing countries is a challenge to the management capabilities of national and local authorities. In his chapter *Toward Gigapolis? From Urban Growth to Evolutionable Medium-sized Cities*, Marco Keiner asks if there are limits to urban growth, and if cities can continue to grow until they become ungovernable 'gigapolises'. This has to be avoided. Sustainable, and more precisely, 'evolutionable' development is a must, and early action is needed. This chapter is devoted to the plight of relatively small but rapidly growing cities and proposes several prerequisites in order to successfully implement and attain sustainable urban development and allow an evolution of socioeconomic and environmental values and capital stocks.

The transfer of technology, knowledge, and experience from the North to the South, from the so-called 'developed countries' to 'developing countries', has been promoted over decades by international institutions as well as national aid agencies. The rationale behind such endeavors was to assist these countries in their efforts to overcome underdevelopment. In this context, the chapter *Urban Planning in the North: Blueprint for the South?* by Klaus R. Kunzmann critically explores the transfer potential of practiced urban planning in the cities of the North to the cities of the South. He argues that concepts of urban planning are quite different depending on the context, and urban problems, urban development instruments, and urban policy differ considerably from North America to Europe, from the Anglo-American tradition of urban planning to traditions in Scandinavia and Germany, Italy or France. Urban development problems in the South, in turn, are also far from being homogenous. Best practice transfer from one country to another is extremely superficial when this is not taken into consideration. The difficulties and traps of this kind of transfer are discussed, such as the vested interests of the development industry in the North, the ideology and paternalism of planners in the North misusing the cities in the South as testing grounds for development theories, and the problematic attitudes of urban bureaucrats in the South and their limited absorptive capacity. The chapter ends on a more positive tone by presenting a few principles under which urban planning transfer could eventually make some sense.

Note

1   http://www.unchs.org/habrdd/global.html (3 December 2004).

PART I
THE CHALLENGES OF
URBAN DEVELOPMENT
PROCESSES

Chapter 1

Asian Urbanization and Local and Global Environmental Challenges

Hans J.A. van Ginkel and Peter J. Marcotullio

Asia has experienced rapid development and urbanization in the last half of the 20th century. The transformations have reached unprecedented scales. Many indicators have been used to demonstrate the improvement in overall human well-being, notably those advancements related to income, longevity and some health measures, but this only tells part of the story. The great transformations now in process have been accompanied by both local and global challenges to environmental sustainable development. Moreover, there are indications that the challenges will increase in intensity into the future. It is crucially important to identify the dynamics underlying these developments and point to policies that could begin to address their impacts.

This chapter describes the economic and urbanization transformations experienced by Asian cities. It then explores some of the local and global environmental challenges that have accompanied these changes. Specifically, for local issues, we identify the 'infrastructure time-bomb' – defined as the overburdening of old, out-of-date, and overused basic infrastructure (e.g., water supply and sanitation) – within some of these cities. Future urbanization will only increase the burden on these structures and if not addressed, the situation may grow into a crisis.

In order to address these issues, integrated approaches to planning and urban management are needed. We look to specific examples within the Netherlands to see how a small country, characterized by high population density, solved its resource and spatial planning challenges. The results suggest some lessons for Asian cities. Finally, these observations and analyses have encouraged the United Nations University (UNU) to rethink the direction of its Urban Programme. The last section presents thoughts on how this new program will provide research and capacity building to help address these issues.

Asian Urbanization and Local Environmental Challenges

Asia has been undergoing a rapid and intensive urban transition. As Table 1.1 demonstrates, since 1950 more than 1 billion people have been moved into the region's cities, bringing the percentage of urban population up from 17 to more than 36 per cent. This immense process of urbanization is far from over. With only 38 per cent of the region's population currently living in dense human settlements,

11

Table 1.1 Comparative aspects of Asian urbanization (population in thousands)

Asia	1950	1960	1970	1980	1990	2000	Change 1950–2000
Total population	1,402,021	1,702,321	2,147,021	2,641,339	3,180,594	3,682,550	2,280,529
Per cent of total world population	55.6%	56.3%	58.1%	59.5%	60.4%	60.8%	
Urban population	244,085	353,895	502,518	709,214	1,029,251	1,351,806	1,107,721
Per cent of total world urban population	32.5%	34.8%	37.0%	40.4%	44.9%	47.5%	
Per cent urban	17.4%	20.8%	23.4%	26.9%	32.4%	36.7%	
Urbanization rate (5 year annual average)	1.8	1.5	1.1	1.8	1.2	1.3	

Notes:
Asia includes 50 economies. They are Afghanistan, Armenia, Azerbaijan, Bahrain, Bangladesh, Bhutan, Brunei Darussalam, Cambodia, China, Hong Kong SAR (China), Cyprus, Democratic People's Republic of Korea, East Timor, Gaza Strip, Georgia, India, Indonesia, Iran (Islamic Republic of), Iraq, Israel, Japan, Jordan, Kazakhstan, Kuwait, Kyrgyzstan, Lao People's Democratic Republic, Lebanon, Macau, Malaysia, Maldives, Mongolia, Myanmar, Nepal, Oman, Pakistan, Philippines, Qatar, Republic of Korea, Saudi Arabia, Singapore, Sri Lanka, Syrian Arab Republic, Tajikistan, Thailand, Turkey, Turkmenistan, United Arab Emirates, Uzbekistan, Viet Nam and Yemen

Source: Calculated from data supplied by the UN (1999) *World Urbanization Prospects: 1999 Revision* New York: United Nations:
File 2: Urban Population by Major Area, Region and Country, 1950–2030 (in thousands) (POP/DB/WUP/Rev.1999/1/F2) and
File 1: Total Population by Major Area, Region and Country, 1950–2030 (in thousands) (POP/DB/WUP/Rev.1999/1/F1)
Data sets in digital form.

the next 30 years will bring even greater changes (see Table 1.2). Approximately 1.25 billion of the total 2 billion inhabitants added to global urban population will end up in cities in Asia by 2030. This will change Asia's urbanization rate from 36 to 53 per cent and increase the region's share of the world's total urban population. It is estimated that by 2030 more than 53 per cent of the total global urban population will be found in Asia.

Asian urbanization has occurred at some of the fastest rates in history. The average increases for the region varied between 1.2 and 1.8 per cent, but this masks the high speed of change in some countries. For example, South Korea's annual urbanization rates were over 3 per cent from 1960 to 1980, and for the years 1965 to 1970 reached 4.6 per cent. Thus, not only does the absolute number of individuals

Table 1.2   Comparison of size and share of urban populations by region, 2000 and 2030 (population in thousands)

	2000 population	Share	2030 population	Share	2000–2030 % change
Total world urban population	2,845,049	100.00	4,889,393	100.00	71.86
Per cent world urban	46.99%		60.27%		
Africa urban population	297,139	10.44	765,709	15.66	157.69
Per cent Africa urban	37.88%		54.46%		
Asia urban population	1,351,806	47.51	2,604,757	53.27	92.69
Per cent Asia urban	36.71%		53.41%		
Europe urban population	544,848	19.15	570,612	11.67	4.73
Per cent Europe urban	74.75%		82.58%		
Latin America and the Caribbean urban population	390,868	13.74	604,002	12.35	54.53
Per cent Latin America and Caribbean urban	75.29%		83.25%		
North America urban population	239,049	8.40	313,663	6.42	31.21
Per cent North America urban	77.20%		84.37%		
Oceania urban population	21,338	0.75	30,650	0.63	43.64
Per cent Oceania urban	70.21%		74.41%		

Source: Calculated from data supplied by the UN (1999) *World Urbanization Prospects 1999 Revision*
New York: United Nations:
File 2: Urban Population by Major Area, Region and Country, 1950–2030 (in thousands) (POP/DB/WUP/Rev.1999/1/F2) and
File 1: Total Population by Major Area, Region and Country, 1950–2030 (in thousands) (POP/DB/WUP/Rev.1999/1/F1)
Data sets in digital form.

Table 1.3 Regional comparison of growth in large urban areas, 1950–2000 (population in thousands)

	1950 Pop.	No.	1960 Pop.	No.	1970 Pop.	No.	1980 Pop.	No.	1990 Pop.	No.	2000 Pop.	No.
Urban areas larger than 1 million	**195,456**	**85**	**297,015**	**115**	**451,852**	**175**	**617,683**	**234**	**850,182**	**326**	**1,114,346**	**405**
Africa	2,410	2	6,363	3	14,764	7	28,359	13	57,951	27	101,279	41
Asia	61,005	31	104,164	43	181,612	72	272,094	101	411,979	156	566,153	204
Europe	69,344	28	92,694	36	116,116	48	134,061	59	145,649	64	149,107	64
Latin America and Caribbean	16,833	7	32,894	12	56,504	18	86,305	24	122,733	38	163,601	48
North America	42,837	15	56,916	19	77,855	28	89,805	34	101,898	36	122,286	42
Oceania	3,027	2	3,984	2	5,001	2	7,059	3	9,972	5	11,920	6
Urban areas with more than 10 million	**12,339**	**1**	**25,140**	**2**	**43,843**	**3**	**75,579**	**5**	**162,863**	**12**	**262,648**	**19**
Africa	0	0	0	0	0	0	0	0	0	0	23,979	2
Asia	0	0	10,976	1	27,652	2	33,593	2	93,957	7	149,861	11
Europe	0	0	0	0	0	0	0	0	0	0	0	0
Latin America and Caribbean	0	0	0	0	0	0	26,385	2	41,394	3	59,028	4
North America	12,339	1	14,164	1	16,191	1	15,601	1	27,512	2	29,780	2
Oceania	0	0	0	0	0	0	0	0	0	0	0	0
Urban areas between 5–10 million	**42,121**	**7**	**80,477**	**12**	**130,064**	**18**	**157,559**	**21**	**153,073**	**21**	**155,043**	**22**
Africa	0	0	0	0	5,333	1	6,852	1	16,314	2	5,064	1
Asia	12,253	2	26,836	4	40,754	6	79,263	11	77,018	11	85,500	12
Europe	24,826	4	28,935	4	36,295	5	36,480	5	37,441	5	38,259	5

Latin America and Caribbean	5,042	1	12,199	2	32,588	4	18,661	2	15,508	2	19,269	3
North America	0	0	12,507	2	15,094	2	16,303	2	6,792	1	6,951	1
Oceania	0	0	0	0	0	0	0	0	0	0	0	0
Urban areas between 1–5 million	**140,996**	**77**	**191,398**	**101**	**277,945**	**154**	**384,545**	**208**	**534,246**	**293**	**696,655**	**364**
Africa	2,410	2	6,363	3	9,431	6	21,507	12	41,637	25	72,236	38
Asia	48,752	29	66,352	38	113,206	64	159,238	88	241,004	138	330,792	181
Europe	44,518	24	63,759	32	79,821	43	97,581	54	108,208	59	110,848	59
Latin America and Caribbean	11,791	6	20,695	10	23,916	14	41,259	20	65,831	33	85,304	41
North America	30,498	14	30,245	16	46,570	25	57,901	31	67,594	33	85,555	39
Oceania	3,027	2	3,984	2	5,001	2	7,059	3	9,972	5	11,920	6

Source: Calculated from data supplied by the UN (1999) *World Urbanization Prospects 1999 Revision*
New York: United Nations:
Part 2: Urban Agglomerations (POP/DB/WUP/Rev.1999/2/F10)
Data sets in digital form.

moving to cities in the region dwarf anything previously experienced but the rates
of increase are also unprecedented.

Another feature of Asian urbanization is the rise of large urban agglomerations.
Table 1.3 demonstrates the rise of these urban agglomerations since the middle of
the 20th century. The total number of large cities (i.e., those greater than 1 million)
climbed from 85 in 1950 to 405 in 2000. The largest per cent increase among the
different sizes of cities was the 'mega-cities' category (i.e., those equal to or greater
than 10 million inhabitants). According to the UN (1999), there was one mega-city
in 1950, but this number increased to 19 by 2000. Of the 19 mega-cities in the
contemporary world, 11 (i.e., 58 per cent) are located in Asia. Indeed, Asia has
dominated over other regions in terms of the emergence of large cities in all size
categories (see Table 1.4).

History has shown, however, that Asia has always been the location of the world's
largest cities. The shift of large cities to North America and Europe during the late
19th and 20th century can be seen as a fluctuation rather than a long-term trend.

**Table 1.4   The world's urban agglomerations, 2001–2015 (>10⁶ inhabitants
in 2001)**

Rank			Population		Population share in agglomeration	
2001	2015	Name	2001	2015	% Total	% Urban
1	1	Tokyo	26.5	27.2	21%	26%
2	4	São Paulo	18.3	21.2	11%	13%
3	6	Mexico City	18.3	20.4	18%	24%
4	7	New York	16.8	17.9	6%	8%
5	3*	Mumbai	16.5	22.6	2%	6%
6	12	Los Angeles	13.3	14.5	5%	6%
7	9	Calcutta	13.3	16.7	1%	5%
8	2*	Dhaka	13.2	22.8	9%	37%
9	5*	Delhi	13.0	20.9	1%	5%
10	13	Shanghai	12.8	13.6	1%	3%
11	14	Buenos Aires	12.1	13.2	32%	37%
12	8*	Jakarta	11.4	17.3	5%	13%
13	20	Osaka	11.0	11.0	9%	11%
14	16	Beijing	10.8	11.7	1%	2%
15	17	Rio de Janeiro	10.8	11.5	6%	8%
16	10*	Karachi	10.4	16.2	7%	21%
17	15*	Metro Manila	10.1	12.6	13%	22%

Of which 11 are in Asia
Note: '*' indicates a move to higher rank

N.B.

10	15	Rhein/Ruhr	12.9	13	16	18
4	7	Rhein/Ruhr/Main/Neckar	18.2	18.3	22	25

Figure 1.1 demonstrates that most of the largest 20 cities in the world at different points in time since 1350 BC have been located within Asia. In early times large African cities were located in Egypt, but in later periods large cities appeared in central and southern Africa. Those of early Europe were located in Greece and Italy, but locations changed later to include more Western European countries. Early large Asian cities were first located in Asia Minor, the Middle East, China and South Asia. Between 600 and 800 AD, Japanese and Southeast Asian cities were included in the list of the world's largest cities. Mexico's Teotihuacan was included for a brief period. It was not until contemporary times that cities in North, Central and South America had surpassed others in the world. While cities in Europe and North American grew to a large size much more rapidly than their predecessors as a result of the Industrial Revolution, the new millennium will once again see the rise of large cities in Asia (see Chandler 1987).

The number of large cities will continue to increase in the medium-term. The UN predicts that by 2015, the number of mega-cities will increase to 23 (see Table 1.5). In the next 15 years, however, it is the middle size large cities (i.e., those from 5 to 10 million inhabitants) that will gain the largest numbers and increase in relative size. It is interesting to note that while the proportion of the urban population is expected to dramatically rise, the proportion of those living in mega-cities is not. Large cities are growing more slowly than others. The UN projects that of the total urban population, the mega-cities will continue to maintain a share of approximately

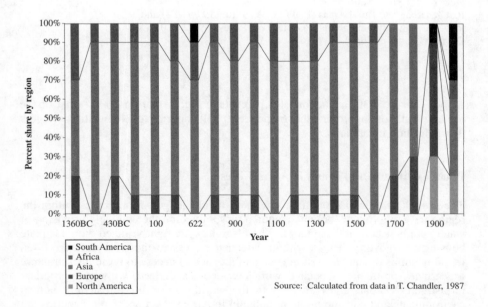

Source: Calculated from data in T. Chandler, 1987

Figure 1.1   Regional shares of world's 20 largest urban areas, 1360 BC to 2000 AD

**Table 1.5   Share comparisons of various sized urban regions, 2000 and 2015
(population in thousands)**

	2000		2030		2000–2015
	Pop.	No.	Pop.	No.	Per cent change
Total world urban population	2,845,049		3,817,292		34.17%
Urban regions of 10 million or more	262,648	19	374,739	23	42.68%
Per cent share	9.23%		9.82%		6.34%
Urban regions of 5 to 10 million	155,043	22	247,839	36	59.85%
Per cent share	5.45%		6.49%		19.14%
Urban regions of 1 to 5 million	696,655	364	866,402	402	24.37%
Per cent share	24.49%		22.70%		
Urban regions of less than 1 million	1,730,703		2,328,312		34.53%
Per cent share	60.83%		60.99%		0.27%

Source: Calculated from data supplied by the UN (1999) *World Urbanization Prospects
1999 Revision*
New York: United Nations:
Part 1: Urban and Rural Areas (POP/DB/WUP/Rev.1999/1/F2) and
Part 2: Urban Agglomerations (POP/DB/WUP/Rev.1999/2/F10)
Data sets in digital form.

12 per cent in 2030, a fraction of a percentage higher than in 2000. Most of the
urban population (approximately 60 per cent) will continue to live in cities with less
than 1 million inhabitants.

Local Environmental Issues in Asian Cities

The growth of cities in Asia is not without its challenges. Many cities within the
region have significant 'brown' agenda issues, including the equitable provision of
water supply access and sanitation. These services are tightly linked with adequate
housing and poverty. Those without adequate access to these services are often
found in slums or squatter settlements and the lack of access to basic infrastructure
exacerbates problems associated with low incomes. Indeed, without appropriate
infrastructure urban areas are among the world's most life-threatening human
environments and half the urban population in Africa, Asia and Latin America are
suffering from one or more of the main diseases associated with inadequate water
and sanitation provision (World Health Organization 1999). Water, sanitation and

hygiene risks are associated with 2.2 million annual deaths and diarrhea alone causes 6000 deaths per day, mostly among children (UN-Habitat 2003).

Local environmental challenges dominate low-income cities (Hardoy *et al.* 2001; McGranahan *et al.* 2001). In Asia the challenges are particularly acute. Table 1.6 presents the numbers of those without access to adequate water supplies and sanitation in Asia. These figures are, indeed, significant. Even though the percentage of those in Asia without adequate infrastructure for these life supporting needs are as high as those in Africa, the absolute numbers in Asia are over three times greater than in Africa or Latin America.

Table 1.6   Estimates for the proportion of people without adequate provision for water and sanitation in urban areas of the developing world

Region	Number and proportion of urban dwellers without adequate provision	
	Water	Sanitation
Africa	100–150 million (ca. 35–50 per cent)	150–180 million (ca. 50–60 per cent)
Asia	500–700 million (ca. 35–50 per cent)	600–800 million (ca. 45–60 per cent)
Latin America and the Caribbean	80–120 million (ca. 20–30 per cent)	100–150 million (ca. 25–40 per cent)

Source: UN-Habitat, 2003

It must be noted that the numbers in Table 1.6 are based upon estimations. Exact comparisons are extremely difficult to make, as there is little agreement as to what constitutes 'adequate' living conditions. At the same time, situations have been documented in individual cities that suggest similar results. For example, in many cities of the rapidly developing Asian region, it is not uncommon that large sections of the population use water that is contaminated and have no provision for sanitation in their home (Hardoy *et al.* 2001). Moreover, even in 'successful' cities, such as Bangalore, almost a third of the population is without direct access to water and must draw water from public fountains (ibid).

Many of these cities are surviving on colonial infrastructure that has been taxed beyond limits. Further, the urbanization within Asia that is expected to continue into the early 21st century is largely within the least developed or low-income economies of Asia. These are those governments that have the fewest resources to provide services. One way to assess the scale of the problem is to match projected challenges against what the international community has set as a goal to overcome. Target 10 of the UN Millennium Development Goal 7 seeks to half the proportion of people without sustainable access to safe drinking water and basic sanitation by 2015, and target 11 seeks to achieve a significant improvement in the lives of at least 100 million slum dwellers (defined as people currently without adequate housing and basic services) by 2020. The latter target falls well short of the 924 million people presently estimated living in slums (Payne and Majale 2004). Between 2000 and 2020, the UN predicts that in low and lower-middle income countries in developing Asia alone there will be an additional 364 and 408 million people added to cities (see Table 1.7). Hence, local environmental challenges are significant and will remain so for the region into the foreseeable future.

Table 1.7   Change in urban population within Asia by income status, 2000–2020

Country status	Countries (n)	Urban population 2000 (in thousands)	Urban population 2020 (in thousands)	Change 2000–2020 (in thousands)
Low income	17	450,262	814,373	364,111
Lower-middle income	17	704,570	1,113,529	408,959
Upper-middle income	4	36,378	61,265	24,887
High income	12	160,599	176,435	15,836

Source: The economic data is from the World Bank. Economies are divided according to 2003 GNI per capita, calculated using the World Bank Atlas method. The groups are: low income, $765 or less; lower middle income, $766–$3,035; upper middle income, $3,036–$9,385; and high income, $9,386 or more. (Based on figures available at http://www.worldbank.org/data/countryclass/countryclass.html in October 2004.)
The data on percentage urban are based on United Nations, 1999: World Urbanization Prospects: The 1999 Revision.
POP/BD/WUP/Rev.1999/1/F2, United Nations, New York.

Asian-style Globalization and Global Environmental Challenges

At the end of the 20th century, the world economy experienced structural adjustments affecting production, resource utilization and wealth creation. Cross-border functional integration of economic activities and growing interdependencies among regional economic blocs are part of a set of processes defined as globalization. Evidence for the geographical scope of globalization usually includes the locations of nodes within the flows, while the intensity of globalization is generally given by a number of trend indicators including trade and financial flows, foreign direct investment (FDI), communications (i.e., information flows) and personal and business travel. The increasing scope and intensity of global economic integration has been underpinned by advances in technologies facilitating communications and transportation.

Globalization has not impacted nations evenly. Those in East and Southeast Asia have experienced the rapid development of a manufacturing belt with high accession rates in secondary economic activity and growth. While most of the world's manufacturing production is still located in OECD countries like the United States, Japan and Western Europe, there have been significant changes in the geography of manufacturing output. In all of the G7 countries, the share of manufacturing in GDP dropped significantly from 1960 to 1993. In the United States from 1979 to 1995, as many as 24.8 million 'blue-collar' jobs were extinguished and new employment was not generated in the manufacturing sector. This decrease in manufacturing employment has translated directly into lower production levels. In terms of world share, the manufacturing output from the USA has declined from 34.4 per cent in 1965 to 25.7 per cent in 1992.

With the loss of manufacturing in many countries of the OECD, an industrial belt has been emerging in the Asia-Pacific region. The Asian Newly Industrialized Economies (NIEs, including Hong Kong, South Korea, Singapore and Taiwan), followed by the ASEAN-4 countries (Indonesia, Malaysia, the Philippines and Thailand) were able to take advantage of the shifting location of industrial manufacturing jobs. The Asian NIEs experienced spectacular manufacturing sector driven growth during the last few decades, while other developing nations did not. The percentage of South Korea's GDP output accounted for by manufacturing increased from approximately 14 to 29 from 1960 to 1993. Singapore's manufacturing component jumped from 12 per cent to 37 per cent of the city-state's GDP during the same period.

The theoretical explanation of patterns of growth among nations in the region has been called the 'flying geese' theory (see, e.g., Akamatsu 1962; Bernard and Ravenhill 1995; Hatch and Yamamura 1996; Kojima 2000; Lo 1994). There are several variants of the theory, although the dominant view combines elements of trade and the product life-cycle related to foreign investments (see, e.g., Vernon 1966) to explain the rapid industrialization process in the region's 'latecomer' economies. The theory starts with single industry countries and explains the diversification and rationalization of their industries as they adopt more efficient production methods. These changes are transmitted through pro-trade-oriented FDI from a 'lead' goose (Japan) to 'follower' geese (NIEs, countries in the Association of Southeast Asian Nations – ASEAN and now China and Vietnam). Development is therefore linked with the life-cycle of emergence, maturation and decline of particular industrial sectors. The variety of levels of development found within the region provides complementary comparative advantages for transnational flows within sectors based upon their life-cycle stage. With Japan in the lead, the other Asian economies follow a trajectory defined by the level of economic development among interlinked and increasingly interdependent nations. The theory portends an avalanche of development as 'latecomers' catch-up. Indeed, one advocate, Kojima (2000: 397) predicts that regional economic development following the 'flying geese' model between all the ASEAN countries, China, Korea and Japan will lead to 'more equal income levels among themselves, say within 20 or 30 years.'

While the theory has been attacked, it retains interest among both academics and practitioners. Importantly, trade and investment flows have been associated with rapid technological diffusion, sometimes more rapidly than expected. China, for example, is developing technology intensive industries (i.e., high-level/capital-intensive activities), which is not expected for a low or lower-middle income country. Approximately 400 of the Fortune 500 firms have invested in over 2000 projects in China, including the world's leading manufacturers of computers, electronics, telecommunication equipment, pharmaceuticals, petrochemicals and power-generating equipment (UNCTAD 2001).

Asian Urbanization and Global Environmental Challenges

At the center of these changes are the cities in Asia. Increasingly, globalization triggers a number of internal changes to cities, as defined by the world city formation process, or the concentration of the world's active capital (see Friedmann 1986; Friedmann and Wolff 1982). The world city formation process has spatial implications at different scales. At the city scale, a number of infrastructure projects and shifts in residential, industrial and commercial location create urban patterns associated with the city's functional role within the regional and global economic system (see, e.g., Marcotullio 2003).

At the metropolitan scale, 'city systems' have emerged including large and small urban centers that have become economically integrated, transcending numerous political divisions and subdivisions at regional, national and sometimes international (e.g., Singapore) levels. This process of extension and integration was first observed by the geographer Jean Gottman (1961), who identified the emergence of the extensive urban region he called 'megalopolis'. He introduced this word to designate the area in the United States, which stretches from Boston to Washington, D.C., along the east coast. The term and concept proved to be so appealing that it was later applied to other extensive urban regions. Subsequently, 25 other megalopolis regions were identified in the USA (Brunn and Williams 1983). Perhaps the second most cited megalopolis is Tokaido, stretching from Tokyo to Hakata/Fukuoka. This megalopolis is particularly evident to anyone traveling on the *shinkansen* from Tokyo to Osaka, for example, as it is difficult for an observer from the train to identify when one city ends and the next begins. The spaces between once bounded cities have been 'filled in' with the advent and application of high speed rail lines, roads and communication networks first, followed by changes in residential, commercial and industrial land use patterns.

McGee (1991) started with this notion and developed the concept of *desakota*, (meaning 'village-city', the words which come from the languages spoken in Indonesia and Malaysia), for the unique large landform associated with urbanization in Southeast Asia. His idea suggests new extended urban activity surrounding the core cities of many countries in the region. The emergence of this particular spatial configuration is the outcome of a unique space-economy transition, unlike that experienced by Western countries. The *desakota* areas are those of intense mixtures of agricultural and non-agricultural activities that often stretch along corridors between large city cores. Subsequent work on this issue has resulted in further elaboration of this compelling idea (see, e.g., Lin 1994; McGee and Robinson 1995).

Others have taken the analysis of the physical connections among cities and projected the possible impact at a global scale and into the future. Doxiadis and Papaioannou (1974) developed the *ecumenopolis* concept in a systematic study of what the 'City of the Future' would include. The *ecumenopolis* is a global city of

30 billion inhabitants, continuously connected around the world through infrastructure networks. These authors maintained that this formation is not only likely – it is inevitable.

At the international scale, the theme of connectivity among cities has been richly described within the Asian region, although not always through direct physical contacts. For example, Choe (1996) provided an illustration of a mature transnational sub-regional urban corridor, in which an inverted S-shaped 1500 km urban belt from Beijing to Tokyo via Pyongyang and Seoul connects 77 cities with more than 200 000 inhabitants each, into an urban conglomeration of more than 97 million people (see Figure 1.2). This formation is discussed as a potential future transnational city system.

Others have envisioned even larger international urban systems. Lo and Yeung (1996, 1998) have suggested that a regional city system is developing in the Asia-Pacific region. This system is based upon flows of goods and services, investments, people and information between the major metropolitan centers, largely situated on the coastlines of countries. It stretches from Japan to Indonesia and then out to Canada and Australia. The regional city system is defined as 'a network of cities that are linked, often in a hierarchical manner based on a given economic or socio-political function at the global or regional level' (Lo and Yeung 1996: 2). The large

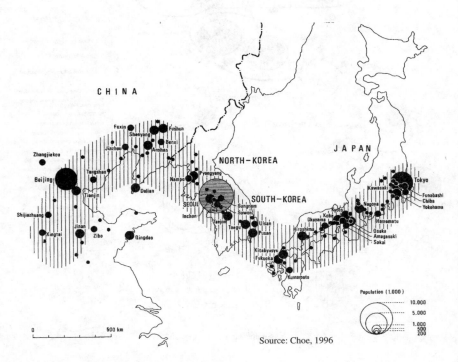

Source: Choe, 1996

Figure 1.2   BESETO urban corridor

urban corridor consists of a set of smaller scale urban corridors, including the Pan-Japan Sea Zone, the Pan-Bohai Zone and the South China Zone, among others (see Figure 1.3).

Globalization, as a driver of urban development has not influenced cities without also impacting urban environmental trends. As trade, investments, people and information have moved throughout the region, so have environmental challenges from disease (e.g., the spread of SARS and AIDS/HIV) to increased consumption

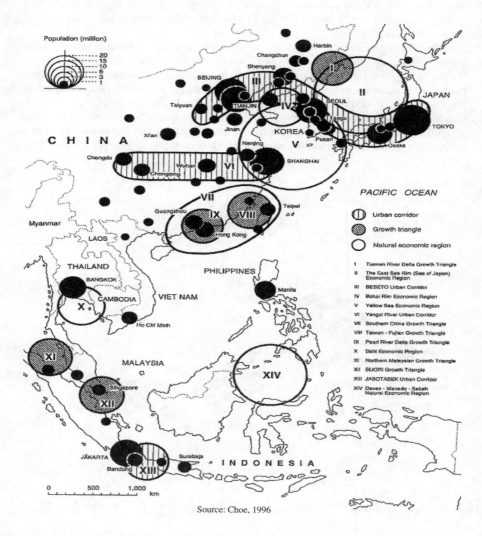

Source: Choe, 1996

Figure 1.3 The Asian regional urban corridor

and waste and emission generation. This is somewhat surprising given current theories on the environmental challenges related to urban activities of low and medium income cities (see, e.g., Hardoy *et al.* 2001). The theory of urban environmental transitions suggests that the types of environmental challenges faced by cities shift with increasing wealth from local traditional burdens to global modern burdens (McGranahan *et al.* 2001). In Asia, it seems that global and environmental burdens are being experienced by countries and cities at lower levels of income per capita (see, e.g., Marcotullio, Chapter 6 in this volume). Globalization-driven growth has fueled the rapidly increasing demand of natural resources. A recent report suggests that China between 1997 and 2002 has stimulated a doubling of its forest product imports, rising from 40.2 million to 95.1 million cbm of round wood equivalents (Sun *et al.* 2004). Import increases are stimulated by both internal and external demand for Chinese low-cost finished wood products (e.g., furniture). The massive increases in the country's timber consumption are fed by forest products in other countries, largely from Russia and Asian economies (ibid.).

Globalization has also helped the diffusion of motor vehicle technologies. As is increasingly evident, Asia has embraced the automobile with a resultant rise in both local and global emissions. In many cities of the region, planners are increasingly concerned with the impacts of the growing car, motorbike and three-wheeled fleets and associated congestion, pollution, noise and accidents. While these issues are local/regional in nature, the internal combustion engine is also an important contributor to global environmental climate change through the emission of greenhouse gasses.

Given the increasing trend in consumption in the region, it is not surprising that Asia's thirst for petroleum is also increasing. What is significant about the Asian experience is the rise in these emissions at increasingly lower levels of urbanization than previously experienced. Figure 1.4 demonstrates that the newest urbanizers are not following patterns set by Japan, South Korea or Hong Kong, all of which produced less CO_2 emissions during specific periods of urbanization than that of the USA. Rather, the newest most rapid developers are increasing their emissions at considerably lower levels in the urbanization process. Moreover, they also emit transportation CO_2 at incomes much lower than those experienced by the West. Table 1.8 compares the levels of transportation related CO_2 levels of the major Asian countries with those of the USA; the major example of a society that stands out among OECD nations due to its high motor vehicle dependency. In Table 1.8, the column for GDP per capita is the level when the emissions were recorded. They are approximately similar among all nations (or lower than that of the USA when it first began to emit these gasses), representing approximate equal levels of development. The years are obviously different and the volumes per capita are significantly different. Table 1.8 demonstrates that for the countries identified, cities are increasingly challenged with issues of global environmental change (see also Marcotullio *et al.* 2004).

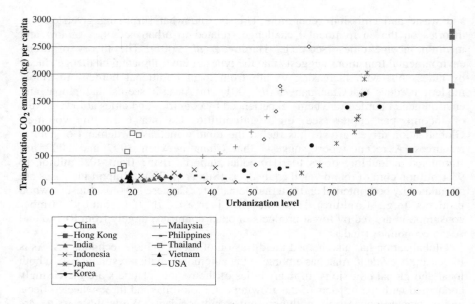

Figure 1.4   Comparison of USA and selected Asian road transportation CO$_2$ emissions (kg) per capita by urbanization level

Urbanization and Planning: The Experience of the Netherlands

Given the complexity of these environmental issues, questions arise as to how to develop adequate responses. While it is difficult to transfer lessons directly from the developed to the developing world, there are some lessons for Asian planning practices that can be gleaned from successful efforts. Here we turn to the Dutch experience with planning and urban policy.

Obsessed with Space

In his interesting analysis *Brilliant Orange: the neurotic genius of Dutch football*, the Brit David Winner (2000) comes to the conclusion that space is the unique defining element of Dutch football: 'No one has ever imagined or structured their play as abstractly, as architecturally, in such a measured fashion as the Dutch. Total football was built on a new theory of flexible space. ... Rinus Michels [the coach] and Johan Cruyff [major player] exploited the capacities of a new breed of players to change the dimensions of the football field' (Winner 2000: 44).

'Total football' was, among other things, a conceptual revolution, based on the idea that the size of any football field was flexible and could be altered by a team playing on it. In fact, it was all about creating space for yourself, destroying space for your opponent and organizing space. This was not an abstract, playful

Table 1.8   Comparison of transportation CO_2 emissions (kg) per capita by fuel type, by GDP per capita and year of selected Asian countries with those of the USA in the year of lowest GDP per capita

Country	From automobiles	From trucks	From airplanes	Total	GDP per capita	Year attained
Japan	161.38	34.58	12.67	208.63	4,426	1961
Korea	42.68	144.53	339.79	527.00	4,557	1982
Hong Kong*	89.92	116.63	352.88	559.43	5,968	1971
Singapore*	321.66	407.36	333.61	1,062.63	4,904	1971
Taiwan	114.06	88.09	180.63	382.78	4,091	1973
Indonesia	131.26	131.22	14.80	277.28	3,655	1997
Malaysia	442.50	258.96	85.51	786.97	4,482	1988
Philippines	102.43	162.98	37.63	303.04	2,363	1997
Thailand	139.65	288.81	95.98	524.44	4,222	1989
China	83.55	29.84	13.61	127.00	3,425	2000
India	20.38	98.29	7.12	125.79	1,910	2000
Vietnam	58.97	111.08	15.42	185.47	1,790	2000
USA	5.02	0.05	0.00	5.07	4,561	1908

* Initial GDP per capita level is higher than that of the USA in 1908.
Source: Marcotullio, P.J., E. Williams and J.D. Marshall (2004) *Sooner, Faster and More Simultaneous*, Report to the USA Environmental Protection Agency (August)

exploration of perspective in the style of an M.C. Escher. Partly, it was instinctive. It was also based on mathematical calculations and pragmatically designed to maximize athletic capacity. In a next stage, time was also integrated into the space concept.

Winner (2000) concludes that this breakthrough of 'total football', based on the concept of space in international soccer, might be one outcome of a much more general obsession of the Dutch with space. Space is scarce and the optimal use of the available space needs the utmost attention. This strong preoccupation is also illustrated by the fact that the Netherlands is widely known for its famous painters and architects, but much less so for its composers or novelists.

It is not so difficult to understand why this is the case:

- The Netherlands is one of the most densely populated countries of the world, that is, 460 inhabitants per sqkm in 2000 (e.g., Belgium: 339, Japan: 321).
- Of the 34 000 sqkm land on which 16 million Netherlanders live, an estimated one third is below sea level, most of which is reclaimed land (i.e., polders).
- The Netherlands are so flat that small 'hills' of 30 meters are called *bergen* (i.e., mountains); as a consequence, the country is characterized by wide horizons, which are put in a framework of long straight canals and artificial lands that have become natural.

- The whole landscape looks like a carefully developed geometric design dominated by cities, in which the original river courses form the major element of irregularity from which, however, long rectangular colonization plots (*copen*) project into the marshy lands.

Quite contrary to the fens of East Anglia, the part of the UK that most closely resembles Holland and where on the ground we see a disorderly countryside divided over and again in very illogical, small portions '… Holland seems a world of order and peace, sense and judgment, where shapes tessellate and the pieces join together neatly' (Winner 2000: 48).

Ruimtelijke Ordening: Interactive Spatial Planning at All Levels

Even though the roots of spatial planning in the Netherlands are much older, systematic and continued spatial planning has started about half a century ago. The first decade after the Second World War was largely devoted to reconstruction. During the 1950s, however, it became evident that some fundamental answers were needed. The most important among these was the solution to the increasing concentration of a growing population in the urbanized western part of the Netherlands, the so-called *Randstad Holland* (rim city; see Figure 1.5) – or 'Greenheart Metropolis' (Burke 1966). This issue was addressed in the official report on *The West of the Netherlands* (1956). Problems were identified both in the West, where the population concentration increased rapidly and scarcity of space became evident, but also in the rest of the country as a consequence of the strong out-migration. It was concluded that a more harmonious development of the country had become highly desirable. From this moment on, the government has developed strategies and policies to achieve a more balanced population distribution.

A key element in this strategy was the regional industrialization policy, which was developed over different stages since 1952. In the beginning, all efforts focused on tackling the problem of structural unemployment (1952–58). In a second stage, the focus was on decentralizing economic growth to diminish the increasing pressure on the western part of the country and to help the North and the South keep their populations (1958–68). In the third stage (1969–72), a major effort was made to create 'counter poles' (or *métropoles d'équilibres*) outside the West, to balance the urban concentrations in North and South Holland, and Utrecht. Increasingly, however, attention was given also to the restructuring of ancient, often derelict, industrial areas. By and large this policy was continued over the following decades, however, with decreasing intensity. Increasingly, attention was also given to the strengthening and renewal of the major cities in the *Randstad*, in particular Amsterdam, Rotterdam and The Hague and, to a lesser extent, Utrecht.

Over time even more important than the regional industrialization policy has been the *ruimtelijke ordenings-beleid* (spatial planning policy) that has been developed systematically since the First Report was published in 1960. The following reports have been prepared and implemented – at least to a certain extent. Their success was summarized by the Belgian president of Grolsch Breweries in the east of the country: '… Each time that I cross the border I see the major difference

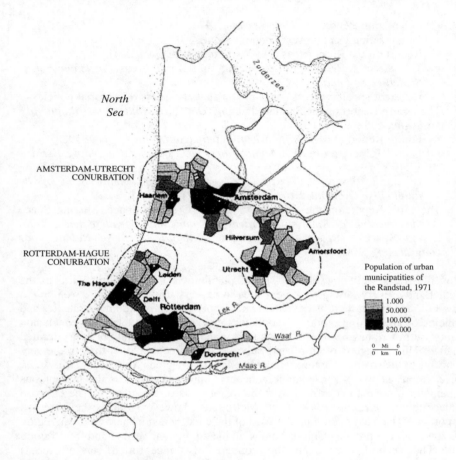

Figure 1.5   Randstad conurbation 1971
Source: Brunn and Williams, 1983.

between our two countries very clearly – The Netherlands has *ruimtelijke ordening*, Belgium not.'

The Spatial Planning Policy Reports were as follows:

1   The Development of the West of the Country (*De ontwikkeling van het Westen des Lands*, 1958)
2   The First Report on the Spatial Planning (*De Eerste Nota over de Ruimtelijke Ordening*, 1960)
3   The Second Report on the Spatial Planning (*De Tweede Nota over de Ruimtelijke Ordening*, 1966)
4   The Third Report on the Spatial Planning (*De Derde Nota over de Ruimtelijke Ordening*, 1974–89)

- Orientation Report, (*Orientatie Nota*, 1974)
- Urbanization Report (*Verstedelings Nota*)
- Report on the Rural Areas (*De Nota Landelijke Gebieden*)
- The Structure Sketch for the Urban and Rural Areas (*Structuur Schets Stedelijke en Landelijke Gebieden*)
- Different specific reports focusing on spatial aspects of sectorial policies
5 The Fourth Report on the Spatial Planning (*De Vierde Nota over de Ruimtelijke Ordening*, 1988–97)
- Fourth Report (*Vierde Nota over ruimtelijke ordening* – VINO, 1988)
- Fourth Report Extra (*Vierde nota over de ruimtelijke ordening Extra* – VINEX, 1992)
- Updating Report of VINEX (*de geactualiseerde Vierde Nota op de Ruimtelijke Ordening Extra* – VINAC, 1997)
6 The Fifth Report on the Spatial Planning – Creating Space, Sharing Space (*Vijfde Nota over de Ruimtelijke Ordening: Ruimte maken, ruimte delen*, 2000)
7 Report Space – Space for Development (*Nota Ruimte – Ruimte voor Ontwikkeling*, 2004)

This illustrates well how crucial spatial planning is in the political life in the Netherlands. Every new government tries to express itself either by developing a new full report or at least by developing a new perspective within the strategies and policies of an existing report. Nevertheless, over the longer term one conclusion stands out: there has been much consistency and continuity and only very limited change. This change is often related rather to the implementation than to the core concepts of the spatial planning. The protection of the 'green heart' of the *Randstad*, for example, is one of those stable concepts. Another concept is the harmonious balancing of population growth over the major parts of the country. The effort to replace the name *Randstad* by 'Delta Metropolis', however, reflects the increased importance given to positioning the West of the country inside the core urban region (or zone) of the European Union. Among all the reports, the second and fourth stood out. The second introduced the concept of concentrated (or clustered) deconcentration; the fourth stressed the compact city concept.

Clustered Deconcentration and Compact Cities

In 1966, the government published the Second Report on Spatial Planning in the Netherlands. In this report, the outline for a society organized along patterns of 'concentrated deconcentration' was elaborated. The idea was to make a clear distinction between urban and rural lands and provide rapid, national transit between the nodes. The plan was enhanced through its elaboration at three levels of government. As time passed, not all the goals were achieved, but there was a noticeable difference in the landscape. By the 1970s, for example, the larger cities in the country – led by Amsterdam and Rotterdam – actively started to reverse deurbanization in favor of compact cities. These urban areas formed the *Randstad*, which was surrounded by green areas, largely under agricultural production. Within the *Randstad*, 46 per cent of the population of the Netherlands resided. The concept was further promoted in the 1985 National Physical Planning Agency report *De compacte stad gewogen* (The Compact City Evaluated), which described the key element of the compact city idea. Essentially the country was striving for an intensive use of space, diversity and multi-functionality. The compact city idea has many theoretical advantages including more efficient energy use, scale economies resulting in lower infrastructure costs and critical mass thresholds that are more likely to justify public transit and non-motorized transport (see, e.g., Capello *et al.* 1999; Haughton and Hunter 1994).

The concept of sustainable development also played a major role in the discussions on the country's urban development (de Roo 2004). The idea reached a popular audience with the publication of the World Commission on Environment and Development (1987) and was subsequently endorsed by the European Commission, which further suggested that the compact city was one way to achieve this trajectory (Breheny and Rookwood 1993). As the compact city notion was already successfully in use, the Dutch government endorsed the idea. Planners were also concerned about a newly arising phenomenon, the decrease in average household sizes, which was responsible for promoting a building program and rapid land use changes (de Roo 2004).

Given the intensive debates over the compact city form, a split developed in the planning literature and among practitioners during the 1990s. Scholars and planners were increasingly skeptical of the compact city idea. In part, dissatisfaction with the compact city idea was related to the conflicts that arose across European cities between environmental policy and spatial planning. Environmental concerns dictated that there should be distance between environmentally intrusive activities – particularly because spatial concentration of environmental pollution brings greater impacts on human well-being – while the compact city idea promoted the opposite. In response, some have argued that the microstructure of cities is as important as its macrostructure in obtaining a form that reduces environmental impacts (Frey 1999). The 'paradox of the compact city' (Breheny and Rookwood 1993) was resolved in the Netherlands through alternative ideas for urban land use patterns that provide micro-level structure advantages. Elements of the alternative ideas included 'clusterd deconcentration' configurations (i.e., multi-nucleated urban areas), transit nodes that integrate the city and its hinterland, and the protection of exurban lands to keep them open and bring nature back into the city (see, e.g., Frey 1999; Haughton and Hunter 1994; Hough 1989).

Reaching the current state of Dutch planning was not a straightforward process. After the Second World War, planners in the country were steeped in the concept of 'rationality', which produced sectoral specialization and resulted in a sharply divided policy system with no way to prioritize outcomes. After searching for various methods to replace the rational problem solving technique, 'integrative planning' was introduced in the 1990s throughout the country through the concept of *omgevingsplanning*, which covers spatial planning, water management, traffic and transport policy and environmental planning. This concept led to the rise of the comprehensive provincial plan. An attractive aspect of the comprehensive plan was the area specific policy approach that transcends sectors, but focuses on the physical environment. In this way it attempts to integrate several policy sectors. It also progresses through an actor consultation model, which enhances public participation in the planning process.

The result of these new planning approaches was the creation of satellite cities, called 'growth cities'. The national railway system was incorporated into the planning effort. The results have been the provision of an excellent service that can move people from major city to major city as easily as it moves them within individual cities.

This experience suggests at least two issues that should be considered in planning for the complex challenges facing Asian cities. First the idea of clusterd deconcentration has merit as a spatial form that will help to provide a high quality of life and lower environmental impacts. Second, the multi-scaled efforts that planners undertook in the Netherlands may provide an idea on how to plan cities in this region. Importantly, cities cannot be managed at the local level alone. One of the important findings of this chapter suggests that larger forces, such as demographic change, globalization and technology diffusion, impact urban environments. Therefore, the planning of cities must be considered at all levels of government and perhaps at the international level as well.

Urban Futures: UNU's New Urban Programme

The challenges discussed in this chapter, along with many other similar trends, have encouraged the UNU to develop a new Urban Programme that addresses the important issues they evoke. The new UNU Urban Programme is motivated by the question: How can we design our urban areas, shape our urban systems, and organize functional urban–non-urban relationships that both enhance growth and promote sustainability in an era of globalization? We perceive studies of each of these areas to take place within three themes: (1) impacts of driving forces of change (e.g., historical, contemporary and predictive studies), (2) traditional and emerging responses to these changes, and (3) visions for sustainable urban futures.

The UN estimates that in the next 30 years the worldwide population share living in urban areas will increase from 48 to 60 per cent. In absolute numbers, the urban population will increase from 2.9 to 5 billion people over this period, that is, an increase of 72.5 per cent (see Table 1.9). Three quarters of global population growth will occur in the urban areas of developing countries, causing hyper-growth in the cities least capable for catering to such increases. The urbanization of poverty, no doubt a partial outcome of this process, is one of the most challenging problems facing the world today and will be into the future. Urban poverty is indeed growing in all cities of the world, but characterizes, in particular, certain aspects of rapidly growing cities in developing countries (UNCHS 2001).

The Core of the Problem

The fact that people are living – and working – in such large numbers, over extensive, contiguous areas in very close proximity, creates many serious problems. A fundamental challenge is to cater for the basic needs of the massive populations involved. To supply sufficient safe water and food, clean air, shelter/housing, employment, health provisions and so on, has often become a gigantic task – in particular in the developing countries. In and around many of the world's urban

Table 1.9   Urban and rural population, 2001–2030 (in billions)

Region	Urban		Rural	
	2001	2030	2001	2030
World	2.9	5.0	3.2	3.3
More developed regions	0.9	1.0	0.2	0.2
Less developed regions	2.0	4.0	2.9	3.1
of which				
Least developed regions	0.2	0.6	0.5	0.7

Region	Per cent urban		Annual growth rate 2000–2005	
	2001	2030	Urban	Rural
World	48%	60%	2.1	0.4
More developed regions	76%	83%	0.4	–0.6
Less developed regions	41%	56%	2.8	0.5
of which				
Least developed regions	26%	44%	4.6	1.8
N.B.				
Japan	79%	85%	0.3	–0.6

agglomerations rapid environmental degradation has become a reality, adding to the already existing human security risks related to the natural environment, such as earthquakes, volcanic eruptions and floods. The ecological footprint of such urban agglomerations reaches further and further beyond the built-up area, and groundwater levels are sinking to alarmingly lower levels. In the cities social problems are increasing as well, that is, inadequate governance, lack of social cohesion, social insecurity, criminality, etc.

To find solutions for these problems is, however, crucial for the future of humankind, to secure a better life in a safer world for all, now and for our grandchildren and their children. Our future will be an urban future. Increasingly the people on earth will live in urban areas of all forms. The question will only be in what kinds of cities we will live. At the same time, the border between the urban and the rural will increasingly become fluid. A decentralization of major urban nodes and a centralization in rural areas will go hand-in-hand, further developing and creating huge urban systems in which the urban and rural will be intertwined.

Concentrated (or Clustered) Deconcentration

Looking at urbanization as the solution, rather than as the problem, we must pay specific attention to two aspects: urban pattern and attitudes, including lifestyles, and preparedness to adopt and apply new technologies, both of which have the potential to promote pathways to sustainability. Rather than experiencing processes of uncontrolled and unhealthy concentrations of population in the biggest (mega-)

cities and a spillover in equally uncontrolled (sub-)urban sprawl, it is worthwhile to make deliberate efforts in developing a more organized process of concentrated (or clustered) deconcentration. Undertaking this effort begins with the definition of major lines for public transport and communication, in order to reduce motorization and to develop different sizes of cities with different characteristics. This is a policy pursued in the Netherlands, relatively successfully. It is also being pursued currently in some parts of China, where we see major uncontrolled movements of rural populations to the big cities. The questions that the Chinese are asking include: How can we develop strategies so that migrants do not all go to Shanghai or Beijing? How can we develop other centers at the lower levels, each with their own specific characteristics, so that we create systems of major cities with satellites, all of which have a different atmosphere – or 'flavor' if you will – defined by different social and environmental characteristics, urban forms, sizes, layouts and functions, creating a unique quality of life for the people living there? How can we generate different kinds of attractiveness or different urban 'milieus'? These questions go beyond national borders and indeed must become more than a strategy pursued nationwide. They must include transborder situations, even international efforts, pursued at different levels of geographical scale simultaneously.

It is clear that reality does not exist on just one level of geographical scale, be it local, regional or national. In fact, all developments at the local level also have a direct impact on the regional or national level and vice versa. Too often we are simply singling out one level without considering the impacts on other levels. Only when we are prepared and able to bring the different levels of horizontal and vertical integration and coordination together, can we successfully identify solutions. It may be understood, however, that relationships exist between the local and the regional level in some aspects only and not in others, whereas the pattern of relations between the local or the regional and the national levels can be different (e.g., subsidiarity).

Form and Function: The Formation

Improved means of transportation and communication enable us to meet face-to-face. The reality is that we can fly from Tokyo to New York within 14 hours. In the 1940s, it took five days to fly from Jakarta to Amsterdam! Whether it is in fashion, sports, culture, economics or politics, we are closer than ever before. On the one hand, information and communication technology replaces physical mobility, but on the other hand, it enhances mobility and face-to-face contacts. The quality of the local environment and the attractiveness and diversity of the local support base are becoming ever more important parts of what is called the (urban-)ecological complex (van Paassen 1962). It is the quality of the local environment, the combination of elements of high quality in administration, industry, trade, science, and culture, which characterizes and structures the city and the agglomeration and gives each of them their specific identity. This characteristic combination of artifacts and institutions and their relations is sometimes called a 'formation'. Formations can be classified in a typology, but will in the end always be individual. To understand the future of our urbanized planet we must develop more insight into the character of urban formations, their relations with (urban-)ecological complexes, the structure of those complexes, and strategies to develop more systematically optimal combinations of diverse cities in mega-urban regions (or megalopolises).

These are some of the concerns that drive the new UNU Urban Programme agenda. While space precludes a more extended discussion of the program, we emphasize that it is currently in the planning stages and is focused on the future, when population growth has considerably slowed down or come to an end, populations 'gray', and globalization continues to set the context for growth (see, e.g., Lutz *et al.* 2004). We suggest that the 21st century will be the urban century and that the sustainable development agenda will only be achieved when we begin to change the role of cities from 'parasites' on the environment to 'promoters' of sustainable development.

Conclusion

We conclude that there are significant local and global environmental problems associated with Asian urbanization, and these trends will only continue. Perhaps they will even get worse due to urbanization and strain on the urban infrastructure. At the global scale, Asian cities are more and more the centers of increasing consumption and waste generation. Many of these challenges have global impacts.

The driving forces for these environmental transitions can be found at every scale level from the global to the local. Within their borders, demographic shifts and urbanization have certainly helped to create challenges. Globalization, too, is implicated in these transformations. As cities continually link to each other beyond their national boundaries, intensive flows are enabled. These linkages have led to the incorporation of new technologies, lifestyles, etc. While there are many positive aspects of these developments, there are also concerns. One example is that of CO_2 emissions, which we predict will continue to grow.

Planning at all levels of governance can help us to address these challenges. Moreover, given the complexity of issues, integrated planning is increasingly in demand. Focusing on the Dutch experience, we note that spatial planning was combined with transportation planning at all levels of government to help address both the local quality of life issues as well as national mobility demands. While this is a single and perhaps unique case, there are some lessons to learn from its results.

Given these trends and possibilities for the future and building on the extensive knowledge generated by the Mega-Cities Programme and its ongoing Urban Ecosystems Projects, the UNU is developing a new Urban Programme that will focus on generating research geared to exploring the complexity of urban environmental challenges in the 21st century and making contributions to sustainable urban futures for generations to come.

Chapter 2

Sustaining Urban Development in Latin America in an Unpredictable World

Alan Gilbert

Twenty years after the depths of the 'lost decade', all is not well in Latin America's cities. If Munck's (2003: 168) comment that 'industries close down and the great cities sink under the weight of pollution and over-crowding' is too pessimistic, the New Economic Model (NEM) is clearly not producing the goods (Bulmer-Thomas 1997; ILO 1995; World Bank 2002). National economic growth rates are low, urban poverty is increasing, inequality in an already unequal region is growing and the modern 'urban' diseases of crime, pollution and congestion appear to be worsening. Competitiveness is the ubiquitous policy watchword, but only a minority of urban economies show signs of winning a space in the world's marketplace.

This chapter considers how cities in Latin America are coping with their problems and examines the kinds of challenges that they need to overcome if they are to improve standards of living in the future.

Demographic Growth

In the 1950s and 1960s, the speed of urban growth was a genuine problem. It was difficult for urban managers to cope with population growth of over 4 per cent per annum. Gradually, however, national fertility rates began to decline, the pace of urban migration slowed and economic deconcentration slowed the growth of the largest cities (Aguilar 1999; Bähr and Wehrhahn 1997; Chant 1999; Diniz 1994, Gilbert 1996, 1998).

The tendency for slower urban growth was accelerated by the debt crisis, which impacted more seriously on the large cities than anywhere else. As unemployment rose and living conditions deteriorated, the word went back to potential migrants in the countryside that now was not a good time to move. In some places, it seemed that more people actually left the cities than arrived. For the first time in decades, most of the largest cities were growing more slowly than urban settlements generally.

By the 1990s, the worst of the economic pain had seemingly passed and Latin America's largest cities seemed poised to expand once again. But conditions in the 1990s were different from those in the 1960s or 1970s. Structural adjustment had opened up the region to the brave new world of global competition. Dependent on how they performed, cities flourished or suffered. While a few, like Santiago and

37

Table 2.1 Urban growth by country and most important city, by decade since the 1950s

Country/city	1950–60	1960–70	1970–80	1980–90	1990–2000
Argentina	3.0	2.2	2.3	1.9	1.7
Buenos Aires	2.9	2.0	1.6	1.1	1.2
Brazil	5.0	5.1	4.3	2.7	2.2
São Paulo	5.3	6.7	4.4	2.0	1.7
Chile	3.9	3.0	2.8	1.8	1.8
Santiago	4.3	3.2	2.8	1.9	1.6
Colombia	4.4	4.3	2.7	2.8	2.5
Bogotá	7.2	5.9	3.0	3.3	2.5
Mexico	4.8	4.7	4.5	2.7	2.0
Mexico City	5.0	5.6	4.2	0.9	1.6
Peru	3.6	4.9	3.5	2.8	2.3
Lima	5.0	5.3	3.7	2.8	2.6
Venezuela	6.1	4.6	3.9	2.5	2.5
Caracas	6.6	4.5	2.0	1.4	n.a.

Sources: 1950–90, UNECLAC/UNCHS 2000, 1990–2000 are mainly estimates from UN World Population Division 2001.

Bogotá regained some of their economic dynamism, many others did not (de Mattos 1999; Gilbert 1998; Iglesias 1992; World Bank 1995).

Emigration also helped slow urban growth in many countries. The combination of local poverty and proximity to the 'Great Society' encouraged millions to move to the United States. By 2000, the Hispanic population in the United States numbered 35.3 million (Logan 2001: 1), approximately 19 per cent of all Mexicans, 16 per cent of all Salvadorans, and 11 per cent of all Cubans and Dominicans were living in the USA.[1] Had the emigrants not moved abroad, most cities in their home countries would have grown much faster.

Economic Growth and Stability

Economic stabilization and restructuring was meant to overcome the problems created by Import Substituting Industrialization (ISI) and prepare the region for a prosperous economic future. Unfortunately, even where serious attempts were made to restructure the economy, the promise of rapid economic growth has rarely been met. Chile and Colombia have had good times, but generally the region experienced slower rates of economic growth in the 1990s than in the bad old days of ISI. Even allowing for the consolation of slowing rates of population growth, the NEM has ushered in a period of economic vigor only in Chile and, to a lesser extent, Colombia.

Table 2.2   Economic growth in selected Latin American countries since 1950 (annual growth in GDP)

Country	1950–59	1960–69	1970–79	1980–89	1990–99	2000–2002
Argentina	2.4	4.4	3.0	–0.6	4.9	–5.5
Brazil	6.5	6.2	8.6	2.9	2.9	2.4
Chile	3.8	4.5	2.0	3.2	7.2	3.1
Colombia	4.7	5.0	5.7	3.7	3.3	1.8
Guatemala	4.0	5.2	5.9	0.9	4.2	2.6
Mexico	5.9	7.1	6.5	2.1	2.7	2.5
Peru	4.9	5.6	4.0	–0.2	5.4	2.6
Venezuela	8.3	5.4	3.2	–0.8	1.7	–0.2
Latin America	4.9	5.7	5.6	1.7	3.4	1.2

Source: United Nations (1998), World Bank (2000), UNECLAC (2003).

What is perhaps more worrying is that the NEM has not solved one of the region's continuing problems: economic instability. One of the major problems with economic integration is the increasing volume of highly mobile capital that is invested in local stocks or bonds. Such portfolio investment moves to wherever the return is highest. When it pours into a potentially profitable country, it forces the exchange rate upwards and makes exporting more difficult. But the real problem comes when it suddenly pours out: when countries fail the latest test of economic competitiveness with Standard & Poor's or even because of events occurring outside the country. The contagious impact of Mexico's 'Tequila Crisis' on the rest of Latin America in 1994 and Brazil's dose of Asian flu in 1994 shows how suddenly, and often unfairly, money managers in London, New York and Tokyo can pull the plug on even the comparatively innocent (Fernández-Arias and Hausmann 2000). The world economy has always been less than stable, but currently the sheer volume of capital that can be transferred at the flick of a switch suggests that it is becoming less stable than ever. If that was not bad enough, the speed of technological change is much faster today. In the past, technology did not change all that much over a thirty-year period. Today, in the fields of electronics, computing and bioengineering, it can change in five years. Plants set up in one place are in constant risk of being made redundant. Enhanced capital flows based on rapidly changing technology do not guarantee stability. Keynesian thought and the establishment of the Bretton Woods institutions were meant to remedy economic instability at a world scale. Today, it is clear that they cannot cope with the increased quantities of capital moving around the globe.

In sum, economic liberalization can bring prosperity to Latin America, but the power of international capital and the speed of technological change mean that, ultimately, much depends on who you are, where you live and whether the sun happens to be shining in New York.

Foreign Earnings Potential

The shift from ISI to NEM was meant to improve Latin America's international competitiveness. But, if we measure the change in export production in terms of its contribution to gross domestic product (GDP), the results for the region as a whole are not all that impressive. While the overall contribution of exports almost doubled between 1965 and 2001, what is fascinating is that so little changed after 1980, the period of structural adjustment and introduction of the NEM. Most of the changes seem to have occurred during the final stages of the period of import substitution. Considering the evidence for the region's seven largest countries shows that there was considerable variation. Two countries, Chile and Mexico, stand out as export 'success stories', Colombia as a partial success, and Argentina, Brazil, Peru and Venezuela as anything but.

What has changed more consistently in Latin America is the form of export production. In the past 20 years, most economies have shifted from an almost total dependence on primary products to a position where many have become significant exporters of manufactures. For the region as a whole, the share of manufactures in total merchandise exports increased from 20 per cent in 1980 to 48 per cent in 2000. Every one of the larger countries shifted its export structure in a similar direction, although the extent of change varies considerably. For example, Chile and Venezuela still concentrate on primary exports whereas Mexico and Brazil now export large quantities of manufactures. The fact that the region's major success story, Chile, produces so few manufactures should be a salutary lesson for current orthodoxy.

The urban consequences of this element of the NEM are fascinating. For while some manufacturing centers gained, many lost out. Whereas ISI favored the major cities, in some places the NEM changed the balance of regional advantage. In Chile new exports were linked to natural resources, and so gave a boost to those regions where the raw material was produced. But in Mexico, the growth of manufacturing exports only helped urban areas located close to the US border. Since most of the manufacturing growth has been confined to *maquiladoras*, the traditional manufacturing centers of inland Mexico have benefited little from the shift. Indeed, when linked to domestic recession and reduced import controls, the NEM brought about a major decline in Mexico City's industrial dominance.

The economic future of Latin American cities increasingly depends upon their ability to stimulate exports. But with rapidly changing technology and fluctuating exchange rates constantly modifying their competitiveness, exporting is a very risky business. As Edwards' (1995: 303) analogy with Alice's observation in *Through the Looking Glass* puts it: 'It takes all the running you can do to keep in the same place. If you want to go somewhere else, you must run at least twice as fast as that.'

One other factor is also becoming vitally important in terms of generating foreign revenues, that is, emigration. On the negative side, large numbers of highly qualified and motivated people have been lost to the local economies by their decision to work outside the region. On the positive side, many of these people are sending money back to the region. In 1995, Mexican migrants sent back US$4 billion of their earnings to their families in Mexico (Castro and Tuirán 2000) and in 2001, a staggering US$9.2 billion (*The Economist* 11 January 2003). In some Central American countries, as much as one quarter of the GDP comes from foreign remittances.

Employment and Unemployment

Until 1980, economic growth was creating lots of jobs in the cities, although formal employment rarely kept pace with the rapidly growing labor force. The result was that the so-called 'informal sector' was growing at the same rate as the formal sector. Unfortunately, the debt crisis of the 1980s increased the contribution of low-productivity services because it badly hit the manufacturing sector (Gilbert 1998; Tardanico and Menjivar-Larín 1997).

During the debt crisis, the contribution of the informal sector to non-agricultural employment in Latin America rose from 40 per cent in 1980 to 54 per cent in 1990, rising especially quickly in the economies most affected by recession (Thomas 1995: 46). Some of those who lost their jobs joined the ranks of the unemployed. But the very poor were forced into the informal sector in order to eke out an existence (Escobar and González de la Rocha 1995).

When economic growth returned in the 1990s, it created more work, but restructuring still led to a rise in unemployment in some places. The numbers out of work rose dramatically in Argentina and Venezuela during the mid-1990s and rocketed suddenly in Colombia and Chile in the late 1990s.

Even when work has been available, one of the big problems has been that so many jobs have been temporary or part-time (Roberts 1994). In Mexico, the heavily unionized plants of central Mexico lost out to the new, generally non-unionized plants of the north. Many corporations managed to reduce the rights of their existing workers. Simultaneously, governments were modifying their labor and social security legislation, increasing the power of the companies over those of their unions (Arriagada 1994). More firms began subcontracting to small producers in the shantytowns and more workers survived on part-time or casual work.

Poverty

Between 1981 and 1989, per capita incomes in Latin America declined by approximately 8 per cent. However, the brunt of the crisis was borne by those living in the cities. For the first time in generations, poverty in the cities increased faster than that in the countryside. While the rural poor increased by 8 per cent between 1980 and 1990, the number of urban poor virtually doubled (see Table 2.3). Since the urban sector had gained most of the benefits from the old model it was not unreasonable that it reaped the withered crops from the ill-invested seeds. Unfortunately, most of the burden was thrust upon the poor; the rich had already converted their funds into dollars and even invested abroad (Portes 1989; Tardanico and Menjívar-Larín 1997).

Structural adjustment in the 1980s was meant to improve the macroeconomic structures of Latin American countries and the benefits should have materialized in the 1990s. Table 2.3 shows that the results were not all that impressive. Nor have matters yet improved greatly in the first years of the new millennium, and, in countries like Argentina, Uruguay and Venezuela, real crises have occurred. The slow pace of economic growth has helped perpetuate the urbanization of poverty.

Table 2.3   The incidence of poverty[2] in Latin America from 1970 to 1999

	Total		Urban		Rural	
	Millions	Per cent	Millions	Per cent	Millions	Per cent
1970	116	40	41	25	75	62
1980	136	41	63	30	73	60
1990	200	48	122	41	79	65
1994	202	46	126	39	76	65
1997	204	44	126	37	78	63
1999	211	44	134	37	77	64

Sources: UNDIESA (1989: 39), Altimir (1994: 11) and UNECLAC (2003).

Since 1990, the incidence of poverty has declined in some countries, notably those in which the economy has achieved respectable rates of growth, but increased in most, most commonly where the economy has been in decline (see Table 2.4).

Table 2.4   Poverty and inequality in Latin America during the 1990s

Country	Year	Gini coefficient	% persons below 50% of mean per capita income		
			Total	Urban	Rural
Brazil	1990	.501	39	52	46
	1999	.542	44	54	47
Chile	1990	.554	54	45	48
	2000	.559	55	46	39
Colombia	1994	.601	49	48	46
	1999	.572	46	46	40
Costa Rica	1990	.438	32	30	28
	1999	.473	36	35	33
Guatemala	1989	.582	48	46	38
	1998	.582	50	43	44
Mexico	1989	.536	44	43	34
	2000	.542	44	39	46
Venezuela	1990	.471	36	34	31
	1999	.498	39	na	na

Note: Argentina, Bolivia and Ecuador have been excluded because data is only available for urban areas and Peru because data is only available for 1997 and 1999.
Source: UNECLAC (2002: table 24).

The NEM, therefore, is not only failing to cut poverty in Latin America in any significant way but is tending to 'urbanize' poverty. In part, this is because people continue to move to the cities from the more impoverished rural areas. But it is also because rising rates of unemployment and burgeoning casual employment are creating new forms of poverty in the urban areas.

Housing and Services

The shortage of housing in most Latin American cities is indisputable, and in 1996, UNECLAC (1996) calculated that the quantitative and qualitative housing deficit in the region amounted to approximately 54 per cent of the existing housing stock. In countries like Peru and Guatemala, the housing deficit was actually larger than the housing stock. However, it is unlikely that the housing problem has deteriorated over the years (Gilbert 2001). Two vital factors have combined to help maintain and in some places even to improve the quality of housing. First, the people's own efforts at building and consolidating their housing has been highly effective in making up for the deficiencies of both the market and the government. Had it been left to the private sector, housing conditions would have been truly awful. Second, the quality of urban shelter has been greatly assisted by continued investment in infrastructure and services. While the capacity of government and/or private providers has varied considerably between cities, there are few signs that servicing has deteriorated in recent years. In most cities, a higher proportion of households have access to electricity and water than ever before.

Of course, the expansion of self-help housing is not without its attendant problems. Building a self-help home takes many years and families suffer while electricity, water, sewerage, transport, health and education services trickle into their settlements. And while the quality of construction is often impressive, the ability of the buildings to withstand 'natural' disasters is being increasingly questioned. Recent disasters like the earthquake in Armenia and the landslides in central Venezuela are testimony to that fact.

Of course, we have the technology today to solve most problems and we certainly know how to prevent natural events from becoming widespread disasters. But, the problem with technology is how it is put to use. In Latin America, it sometimes produces new world trade centers and does not reach the shantytowns.

Nowhere are the contradictory achievements of technological advance better demonstrated than in the arena of transport provision. Although technological advance has offered the promise of rapid movement, in many cities rush hour journeys now take much longer. The key problem is that car ownership has been growing too fast. Road capacity has increased but much more slowly than the number of cars. The consequences are obvious: traffic congestion worsens and the quality of the air deteriorates.

Rather than seriously controlling use of the private car, technological advance has been used to develop capital-intensive 'solutions'. While 'metros' or subways are very popular locally and help more people to get to work, they do not reduce traffic congestion for long. Few governments have been brave enough to eschew expensive metros for much cheaper dedicated bus routes. The future of transport, and infrastructure provision generally, depends on how many governments are prepared to replicate the examples of Curitiba, Brazil, and Bogotá, Colombia.

Urban Governance and Democracy

One positive feature of Latin American cities is undeniable: they are more democratic than they have been for years. Unlike the 1970s, most national governments are now freely elected. Those democratic governments have generally encouraged decentralization, and today, most mayors and town councils are also freely elected (Campbell 2003; Myers *et al.* 2002).

While the benefits of local democracy have yet to be empirically validated, especially in an environment of neoliberal management, privatization and growing belief in the value of public-private partnerships (Gilbert 1992; Batley 1996), some cities are managed much better than they once were. Curitiba, Brazil, is the epitome of this transformation with its enlightened leadership and excellent public transport system. Porto Alegre, Brazil, is another municipal leader with its introduction of social budgeting (Gret and Sintomer 2004). And Bogotá, the capital of Colombia, has been transformed from a city facing a real crisis of governance in the mid-1980s to one where service delivery has improved, the finances sanitized and genuine pride in the city has been generated (Castro 1994; Gilbert and Dávila 2002). In Bogotá, a series of democratically elected mayors after 1988 transformed the way the city's government operated. In combination, these mayors put the public finances in order, invested in substantial public works programs, improved the transport infrastructure, and upgraded many low-income settlements.

Whether decentralization and local democracy can produce similar kinds of success in the majority of Latin American cities is a fascinating question. The position of the smaller cities is particularly pertinent here because most of the region's urban growth is occurring in cities with less than one million inhabitants. Since many of these cities are very short of financial resources and management skills, this is a critical issue. At its best, decentralization may free cities from the clawing controls of national governments. At its worst, it may leave the local population without adequate infrastructure and services and the economy with limited ability to compete internationally.

Crime and the Social Fabric

One of the fears long expressed about cities is that urbanization encourages criminality and current wisdom is that crime is getting out of control. Rotker (2002: 8) observes that: 'It is estimated that each "Venezuelan between the age of eighteen and sixty will be the victim of seventeen crimes, four of which will be violent."'[3] Anecdotal evidence from Caracas, Rio de Janeiro, São Paulo and Mexico City seems to tell a similar story: there is more crime and more of that crime is violent.

However, the problem of crime is often exaggerated. First, crime rates vary considerably across the region, even in the same country. In Colombia, for example, the number of murders per 100 000 people in 2002 ranged from 225 in Medellín to 51 in Bogotá (Bogotá 2002). In Santiago de Chile, very few people died violent deaths. Second, crime rates are not rising everywhere. In Mexico City, rates of theft have risen since the 1980s but murder and violent crime have decreased (Picatto 2003). In Bogotá, too, there has been a clear fall in the homicide rate and in some kinds of crime since the early 1990s. Finally, we have to be extremely circumspect about the reliability of the figures. Few crimes are generally reported because so few people have insurance cover and few trust the police to capture the criminal.

Insofar as crime has risen in Latin America over the last 15 years, the relationship with urbanization is unclear. Normally, some other factor confuses the causative link between urban growth and crime; some authorities blame growing poverty, others the modern demand for material goods. The only clear explanation is where drug use is on the increase: the most violent cities in the region tend to be those where drug gangs are competing over turf.

Conclusion

The challenges facing the cities of Latin America are considerable. Although the metropolitan centers are now growing relatively slowly, at least in comparison with the days of rapid city-ward migration in the 1950s and 1960s, they face a major challenge in terms of overcoming poverty, generating employment, providing adequate infrastructure and services, and solving the problems associated with growing inequality. Urban governments have to address the problems associated with poverty, but they also have to solve the same kinds of problems that face all modernizing cities, the diseases of traffic congestion, environmental degradation, inequality and crime. Some have a decent chance of succeeding, but in an increasingly competitive world, the majority will probably fail.

The external economic context in which Latin American cities must develop is not auspicious. Under existing world trade conditions, urban areas are expected to compete against one another on a playing field tilted in favor of the developed countries. To succeed, cities will have to be well-governed, produce exports and be located in countries that are stable economically and politically. This is a taxing list of requirements and, while some cities will succeed, others will clearly fail. The policy of the United States and Europe with respect to trade is extremely important.

Inequality will grow in virtually every city and the incidence of poverty will increase in those cities that fail to clear the hurdles of greater competitiveness. Growing inequality and poverty may eat at the roots of democratic government and may lead to rising political and social instability. It may also encourage even more people to move within the Americas. The United States will continue to act as a magnet for Latin Americans disappointed with the opportunities that are on offer at home. It will also act as a role model for the people of the region and a threat to those governments who do not perform as the US expects (Munck 2003).

Few cities in Latin America have suffered from urban anarchy in the last few decades and some have become better places in which to live. Variability is, and always has been, the pattern. In the past, most people in Buenos Aires, Argentina, or Montevideo, Uruguay, had a much higher standard of living than most in Managua, Nicaragua, or La Paz, Bolivia. Today, Santiago, Chile and Pôrto Alegre, Brazil, are more obvious paragons; while other cities in Central America, the Andes, or Northern Brazil jostle one another for the wooden spoon. Since variability is the pattern, it would be foolish to offer any kind of sweeping generalization for the future of Latin American cities. Given that the neoliberal agenda now means that cities across the globe are in fierce competition with one another, how different Latin American cities manage to perform will determine their economic prospects. The economic record will help determine the incidence of poverty. And although the NEM is likely to guarantee that most cities will become more unequal, the future of absolute poverty is much less certain.

Notes

1   The number of migrants was divided by the combined total of the current populations of these countries plus those in the United States. Of course, this underestimates the impact in the sending countries because so many migrants are undocumented and because increasing numbers are now migrating to Europe and other international destinations (Jokisch and Pribilisky 2002; Gilbert and Koser 2002).

2   The measure of poverty used in the table is similar to that which is employed in the European Union. As such, it is as much a measure of inequality as one of poverty. For example, if the rich should get richer, faster than the poor get richer, the incidence of poverty increases – seemingly a perverse conclusion.

3   To put this in context, in England over a similar period I have lost six bicycles, suffered three break-ins, lost one car (temporarily) and had been attacked once in the street (on my bicycle due to road rage) – of course, I have always lived in safe areas. My accumulated four-year record in Colombia over 30 years is much less benign: one armed hold-up, three robberies from a car, four attempts to pick my pocket (50 per cent success rate), and three unsuccessful attempts by criminals to impersonate police searching for forged dollars.

Chapter 3

The Urban Challenge in Africa

Carole Rakodi

Sub-Saharan African towns and cities are often portrayed as being in crisis. The challenges that they face include rapid population growth, unaccompanied by industrialization or economic growth; lack of economic dynamism; governance failures; severe infrastructure and service deficiencies; inadequate land administration; and poverty and social breakdown. However, urban centers continue to grow and function despite the severity of these challenges, a phenomenon requiring a closer look.

Inevitably, the fortunes of towns and cities are bound up with the fortunes of the countries in which they are located, and the prospects for some African countries involved in war, with minimal economic resources, or with authoritarian regimes are very poor. For many countries, the future is highly uncertain, given their adverse incorporation into the world economy, their limited autonomy, and the difficulties their incipient democracies face attaining stability. In other African countries, however, there are promising developments with respect to the consolidation of democracy, adaptation of democratic forms to local circumstances, economic recovery, public sector reforms including decentralization and the emergence of stronger civil society. Although there seem to be governance failures, severe infrastructure deficiencies and inadequate land administration systems in many towns and cities, other aspects of urban life, such as family and kinship structures, cultural and associational life, and indigenous social organizations demonstrate considerable resilience. In addition, there is considerable potential for learning from and building on evolving informal systems and the interfaces that are developing between them and formal systems.

The Challenges of African Cities

Sub-Saharan African cities are often portrayed as being in crisis. The challenges that they face are well-known:

Rapid Population Growth, Unaccompanied by Industrialization or Economic Growth

A third of sub-Saharan Africa's population was estimated to be urban in 2000 (34 per cent, about 210 million people), up from 13 per cent in 1950 and 23 per cent in 1980 (UN-HABITAT 2002). It is expected that the urban population will grow to

46 per cent of the total by 2020. Although most governments aimed to industrialize in the years after independence, most made only limited progress, and the import substitution model that was adopted proved costly and limited. Marginalization in the world economy, deteriorating terms of trade for primary products, limited availability of domestic capital, failure to attract foreign direct investment, increasing indebtedness, wars and natural disasters, limited government capacity and economic mismanagement contributed to low or negative rates of economic growth for most of the countries most of the time since the 1960s – and especially in the late 1970s and 1980s. The average African country's urban population grew by 5.2 per cent per annum over the 1970–95 period, while its GDP per capita was falling at an annual rate of 0.66 per cent (Fay and Opal 2000).

Lack of Economic Dynamism

African towns and cities, as a result, have economies that cannot support their growing populations. 'Cities in Africa [the World Bank asserts] are not serving as engines of economic growth and structural transformation. Instead they are part of the cause and a major symptom of the economic and social crises that have enveloped the continent' (World Bank 2000a).

Limited investment in the secondary and tertiary sectors of national economies at the time of independence led governments to invest heavily themselves in direct production, sometimes by nationalizing existing privately owned concerns, and try to attract foreign investment. Concerns over the inherited domination of some economic sectors by non-indigenous – generally Asian – populations led to restrictions on who was permitted to enter certain areas of economic activity, and in the most extreme case, to the expulsion of an Asian population (from Uganda). However, limited domestic and even more limited external investment has constrained the size of national and urban economies. Many urban workers, in some urban centers the majority, have no choice but to engage in informal sector activities, many of them extremely small-scale and with very low profit margins.

Governance Failures

The fortunes of towns and cities cannot be detached from the demographic patterns, economic trajectories and governance arrangements of the national spaces in which they are located. However, to rapid population growth, economic stagnation, a 'chequered' political history, and limited administrative capacity at the national level must be added local governance failures. These can be attributed to inherited centralized administrative systems, post-independence centralization, and the lack of resources and capacity at the local level. As a result of the unfunded mandate of municipal authorities, shaky democratic arrangements, undeveloped civil society organizations, a lack of local administrative capacity and poor coordination between public sector bodies, there has been limited capacity to cope with urban growth or stimulate economic development. The ineffectiveness of local governments has reduced their ability to raise resources or attract political interest and commitment from local residents and politicians, fuelling the vicious circle of weak and unaccountable administrations.

Severe Infrastructure and Service Deficiencies

As a result of the low per capita GDP of most African countries and their limited governance capacity at all levels, infrastructure and service provision have been unable to keep pace with rapid urban growth. Infrastructure installation has been inadequate and inequitable, utilities and social services have not been provided in advance of new settlement or in many newly developed areas, and service delivery is unreliable due to poor maintenance, low charges in comparison with operating costs, and often, limited technical and administrative expertise. As a result, investment is deterred, and residents are subject to increased health hazards and adverse environmental impacts as density increases, including polluted groundwater, rivers and coastal waters as a result of untreated sewage and leaching from poorly managed solid waste disposal.

Inadequate Land Administration

In most countries, indigenous land administration systems were considered inappropriate for urban areas where property ownership was confined to Europeans and the need for planned urban development was acknowledged. However, the imported systems based on European legal principles, concepts of tenure and administrative arrangements have failed to cope with rapid urban growth at low-income levels. As urban centers continued to grow, alternative arrangements for land delivery have evolved. The formal systems have failed to adapt to the demands of the situation, while the informal systems have weaknesses as well as strengths. In particular, unplanned development and lack of clarity and security with respect to tenure typify most African towns and cities.

Poverty and Social Breakdown

About half of sub-Saharan Africa's population is poor – on average, 48 per cent of the population were living on less than a dollar a day between 1987 and 1998. Other indicators of well-being have also deteriorated. The mortality rate increased from 107 out of 100 000 in 1970 to 165 out of 100 000 in 2001, while AIDS is expected to reduce life expectancy at birth by up to 20 years in the worse affected countries. The main triggers of increasing poverty in the urban centers are the loss of wage employment and poor health. Estimates of the extent of poverty vary widely, depending on the definitions and analytical methods used. Generally, analyses using Living Standards Monitoring Surveys or Demographic and Health Surveys show lower incidence of poverty and higher living standards in urban than in rural areas (see, e.g., Sahn and Stifel 2003). However, serious questions can be raised about the methods and figures used, leading some analysts to suggest that urban poverty is greatly underestimated by most conventional statistics and analysis approaches (Satterthwaite 2003). There is considerable evidence that urban poverty worsened in the 1980s and 1990s, as structural adjustment policies, quite deliberately, set out to reduce perceived urban bias. Price decontrol, increased food prices and increased cost recovery for services such as water and electricity hurt the urban poor, already relegated to living in the worst residential areas, worse than most other income groups because the greater part of their incomes was taken up with meeting basic needs.

Urbanization has commonly alarmed social commentators, because of its perceived disruptive effect on established social structures and organizations and the political volatility with which it is often associated. Africa is no exception. It is often asserted that social organizations such as kinship networks are breaking down, harming the socialization of children into prevailing social norms and family support networks. There is less social capital in towns and cities than in villages, so it is said. Further, the political volatility of the urban masses has been advanced as a key explanation of the 'urban bias' felt to have had such damaging effects on both national economies and the well-being of peasant farmers in the 1960s and 1970s.

Certainly, it is true that sub-Saharan African cities and towns face all these challenges. While they can seem so severe that observers panic and despair, the fact that even the largest and most disorderly cities, such as Lagos or Kinshasa (Abiodun 1997; Piermay 1997; Nlandu 2002), continue to function, absorb their growing populations and provide locations for economic and administrative activity must lead us to question whether the crisis is as bad as it is often portrayed. In this chapter, I will consider each of the challenges sketched above in greater depth to ascertain how municipalities and their inhabitants are dealing with the problems. Both the official approaches to administering urban centers and the alternatives that have developed in some cases have strengths and weaknesses, both of which will be considered in order to identify pointers to the future.

Rapid Demographic Growth

The World Bank asserts that Africa's pattern of 'urbanization without growth' is in part due to an urban bias that allowed city dwellers to take advantage of subsidies, such as food pricing and trade policies, favoring urban consumers over rural producers (World Bank 2000a). In addition, worsening physical and economic security in rural areas, wars, and civil unrest in countries like Angola, Liberia and Mozambique have also spurred on the migration of millions to the safety of the cities. In Mauritania, for example, Nouakchott's population doubled during one drought year in the mid-1980s (World Bank 2000a).

Let us examine each of these explanations of rapid urbanization. Fay and Opal carried out a cross-national regression of data from about 100 developing countries between 1965 and 1995. They found that, as many previous analyses have pointed out, urbanization levels are closely correlated to but cannot be explained solely on the basis of levels of income:

> ... changes in income do not explain changes in urbanization. Urbanization continues even during periods of negative growth, carried by its own momentum, largely a function of urbanization... Factors other than income that help predict differences in levels of urbanization across countries include: income structure, education, rural-urban wage differentials, ethnic tensions and civil disturbances. ...' Factors other than the initial urbanization level that help explain the speed of urbanization include: the sector from which income growth is derived, ethnic tensions, civil disturbances and democracy (the

latter two contribute to slowing down the pace of urbanization, all else being constant). Rural-urban wage differentials ... are also significant determinants of the rate of growth of urbanization. (Fay and Opal 2000)

They also assessed potential factors explaining rural out-migration and concluded that there is no systematic evidence to prove that either access to better services in urban areas or better transport means triggers it (Fay and Opal 2000).

Focusing on Africa, Fay and Opal (2000) noted that:

> Africa, at the end of the colonial period, was underurbanized given its income and economic structure, due to the policies of the colonial powers [see also Stren and Halfani 2001]. The 1960–80 period was characterized by very rapid urbanization, even more rapid than can be explained by a catch-up hypothesis, traditional urban-bias measures, agricultural shocks, or civil disturbances. It is largely explained, however, by rural-urban wage differentials. ... [Yet, one] can question whether the particularly high rates of rural-urban wage differentials in Africa in the post-colonial period are a symptom of urban bias. It is possible, but by no means certain. They may have reflected differences in productivity – the fact that they gradually decreased over time at about the same rhythm as in East and South Asia supports this hypothesis.

Price distortions were reduced dramatically as a result of liberalization policies in the 1980s and 1990s, and although they still exist (as do expenditure biases), it is very difficult to assess the magnitude of their impact (i.e., investment levels which exceed the economies of scale/agglomeration and their results in terms of higher returns on those investments). Current levels of urbanization are still more than expected, given Africa's income level, but this cannot, Fay and Opal (2000) suggest, be explained by either rural-urban wage differentials or urban bias. Nor can it be easily explained by conflict and natural disasters, since much of the migration that ensues is short term rather than permanent.

Instead, economic explanations have to be combined with demographic factors. High fertility and declining infant mortality between the 1950s and the 1970s increased the rate of population increase and shifted the population structure between 1965 and 1980 towards the age groups with the highest propensity to migrate. As a result, fertility was initially high amongst urban populations. However, the fertility of urban women is generally lower than that of rural women and, in the African context, evidence is emerging of recent dramatic declines. Demographic and Health Surveys (DHS) show that in Sub-Saharan Africa total urban fertility is now around 5.08 compared to rural fertility of 6.50 (Montgomery *et al.* 2004). Agyei-Mensah (2002) sees signs across a wide range of settings that urban fertility is declining precipitously and observes the most dramatic declines in capital cities. Total Fertility Rates (TFRs) have fallen below 3.0 children per woman in Accra (2.66 in 1998), Nairobi (2.61 in 1998), Lomé (2.91 in 1998), and Harare (2.98 in 1999), and the TFR is rapidly approaching that level in Yaoundé (3.1 in 1998, down from 4.4 in 1991).

> In a careful study in Addis Ababa, Yitna (2002) documents a decline in TFRs from 5.26 children per woman in 1974–99 to 3.17 a decade later, further to 2.2 children in 1989–94,

and further still to 1.76 children in 1998. The largest declines took place in teenage fertility, and these are attributable to a postponement of marriage. (Montgomery *et al.* 2004)

Disaggregated analysis shows that the fertility decline is greatest among higher income women (TFR 85 per cent of the rural level), who are more likely to use contraception, and least among women from the bottom quartile (TFR 92 per cent of the rural level). This is borne out by a study on Nairobi, which revealed that total fertility in Nairobi's informal settlements is 4.0 compared to 2.6 for Nairobi as a whole, 3.5 for other cities in Kenya and 5.2 in rural areas (APHRC 2002). Although HIV/AIDS may also be having an impact, there does not appear to be any evidence for this yet. Thus, while no direct links between macroeconomic indicators and fertility can be demonstrated, the effects of economic crisis and the disproportionately adverse effects of structural adjustment policies on urban populations – such as the growth of youth unemployment – may be leading to indirect effects such as urban housing shortages and the postponement of marriage, which strengthen a general tendency for lower urban fertility.

The initial rise in urban population growth rates due to in-migration and high fertility means that urban areas will continue to grow for some time after fertility has declined, even if economic factors cease to encourage it, perhaps over a period of 20 years. However, the rising share of the urban population in the national total will then have a negative impact on net migration, in addition to any effects of economic crises (Becker and Grewe 1996; see also Potts 1997). Potts (1997) cites evidence of reduced net in-migration and net out-migration in the 1980s and early 1990s. 'The city is somewhat less attractive as the cost of living has risen, subsidies have been removed, services reduced and employment opportunities lessened' (Riddell 1997). As a result of economic restructuring, capital flight, reduced quality of life and increased inequality, 'Many of the African urban poor are now in a worse position than the rural poor and rural-to-urban migration has slowed down considerably, and even reversed in some cases' (Hall and Pfeiffer 2000; see also Simon 1997).

Evidence on the rates and direction of change in urban populations in Africa is unreliable and controversial (Rakodi 1997; Simon 1997). It is almost certainly the case that the proportion of Africa's population living in urban areas, as suggested above, will continue to increase, not least because urban populations will continue to grow as a result of natural increase and boundary extensions absorbing additional populations previously classified as rural. However, the rates of increase in urbanization and urban population growth are, on the basis of the above arguments, likely to decrease. Recent evidence bears this out.

UN figures show that overall, peak urban growth rates were reached in the late 1970s in East Africa (5.36 per cent p.a.), the late 1960s in Middle Africa (5.77 per cent p.a.), the early 1950s in West Africa (5.78 per cent p.a.) and the early 1990s in Southern Africa (3.50 per cent p.a.). In all regions, the rate of growth in the late 1990s was lower than at any time in the past fifty years (see Table 3.1). National rates of natural increase have also slowed – from peaks of 3.04 per cent p.a. in East and 3.05 per cent p.a. in Middle Africa in the late 1980s, 2.64 per cent p.a. in Southern Africa in the late 1960s and 2.95 per cent p.a. in West Africa in the early

1980s (UN 2001). Although there is still net rural-urban migration, over half of urban population growth is now due to natural increase.

Table 3.1   Urban population growth in Sub-Saharan Africa 1995–2000

	Urban growth rate (per cent p.a.)	1995–2000 National rate of population increase (per cent p.a.)	Contribution of natural increase to urban growth percentage (approx.)
East Africa	5.10	2.67	52
Middle Africa	3.97	2.61	65
Southern Africa	3.15	1.61	51
West Africa	4.54	2.67	59

Source: UN 2001.

'The prime determinant of whether urbanization increases rapidly or not is not whether growth is negative or positive. Rather, it is whether the level of urbanization in the country is high' (Fay and Opal 2002). Rates of urban growth and urbanization can be expected to be high when initial levels of urbanization are low. In Africa, these high rates were exacerbated in the early years after independence by urban-rural wage differentials and the high propensity of women anxious to migrate to join their husbands once colonial restrictions on migration were lifted and a youthful population who contrasted lack of agricultural and rural opportunities with growing urban opportunities. Migration, straddling and remittances reduce the risks faced by agricultural households and help to smooth their consumption (Bryceson and Jamal 1997). As the proportion of the population living in urban areas increases, the rate of urbanization slows down for a variety of reasons, although in Africa this slowing down has been counteracted by limited fertility declines, until relatively recently, and the effects of war and disaster (Becker and Grewe 1996; Fay and Opal 2002; Montgomery *et al.* 2004).

Thus the forecast growth rates and populations for Africa's largest cities have not yet come to pass. The growth rate of larger cities is slowing; of 34 cities with populations of more than 750 000, none grew at more than 3 per cent between 1995 and 2000, and 18 were estimated to have grown at less than 2 per cent, or less than the rate of natural increase. Of course, this may in part be an artifact of the failure of boundary extensions to keep pace with urban sprawl.

Smaller cities and towns, in contrast, are growing more rapidly (UN 2001). Arguably, this is desirable since the growth of smaller cities may counter primacy and produce a more balanced urban hierarchy. It may mean that towns and secondary cities are fulfilling economic functions in their surrounding regions and providing intervening opportunities for migrants. Their potential to support the development process is greater if economic institutions such as marketing boards, banks and agro-processing industries do not bypass them. The replacement of state marketing boards with private traders, for example, may encourage economic development in secondary urban centers. Although conditions with respect to service delivery are often worse in secondary centers, the problems are arguably less intractable. Property markets are less overheated than in capital cities and their politics are less volatile because most national politicians live elsewhere, perhaps providing space for the development of a locally rooted and accountable political identity and experiments with more appropriate approaches to urban development.

Economic Marginalization and Lack of Dynamism

There has been little GDP growth in most African countries and what there has been can largely be attributed to the growth of selected parts of the agricultural and extractive sectors. Thus although total GDP grew at 2.3 per cent on average between 1990 and 1998, per capita GDP fell by –0.4 per cent and real incomes have fallen at –1 per cent since the 1980s (UNDP 2000). By the end of the 1990s, per capita output was, in real terms, below that at the end of the 1960s, because of drought and conflict, the adverse terms of the sub-continent's integration into the world economy, and economic and political mismanagement (UN-HABITAT 2003).

Some limited progress was made with developing manufacturing after independence in some countries. However, industry contributed less than a third of GDP overall in 1998 (30 per cent), with manufacturing accounting for a relatively small proportion (less than 10 per cent in half the countries and under 20 per cent in all except Mauritius). In 42 Sub-Saharan African countries (the exceptions being Senegal and Swaziland), manufacturing contributed a stagnant or declining proportion of GDP between 1990 and 2000, despite the hopes that structural adjustment policies would pave the way for the development of export-oriented manufacturing (UNDP 2000). Trade liberalization, the reduction of industrial protection and privatization or closure of state-owned enterprises led to widespread deindustrialization and the loss of industrial employment. The remaining industries have found it difficult to compete with imports or in export markets and little new manufacturing investment has been generated (UN-HABITAT 2003).

Thus, one might allege that the 'industrialization baby' was thrown out with the Import Substitution Industrialization bathwater. Import Substitution Industrialization (ISI) policies were costly and hard to reform once vested interests became entrenched. However, it must be remembered that much industry was taken into or developed while in public ownership because of the absence of any large-scale private capital in most countries at independence, and the problems of ISI cannot be disentangled from those of public sector engagement in direct production. It was not surprising that trade and price liberalization led to the collapse of some industries (such as vehicle assembly in most smaller countries), since they were established for reasons of national pride, a desire for economic autonomy and fashion, and were unviable and unprofitable from the outset and unsustainable in the longer term. However, other ISIs, including those recommended as infant industries for industrializing countries, were unable to adjust sufficiently or cope quickly enough with the massive overnight growth in competition from imports. As a result, even food processing and textiles industries collapsed in the aftermath of structural adjustment.

Sub-Saharan Africa's share of world trade has fallen to under 2 per cent (UN-HABITAT 2003). Although exports of some minerals (e.g., oil, diamonds) and agricultural products (e.g., vegetables, cut flowers) have been sustained or increased, there is little local processing and other primary products (e.g., gold, copper, tea, coffee, cocoa, beef, sugar) have continued to struggle in the face of low world prices and subsidized Northern agriculture. Although the total flow of foreign direct investment (FDI) increased in the late 1990s, less than 4 per cent of flows to developing countries went to Africa (UNDP 2000). In 1997, eight countries experienced net outflows of FDI and 22 more only attracted between US$1 and 2 per capita, mostly in mining and cash crop production with few local economic linkages. Africa continues to be highly indebted, with debt equivalent to over 80 per cent of GDP in 1997 (UNCHS 2001).

Thus, there has been little change in the trends noted in Rakodi (1997) and Simon (1997) in the last decade, despite the much vaunted structural adjustment, debt relief, and ending of some major conflicts.

The continent has been largely unable to transcend its traditional functions in the world economy as a supplier of raw materials and a captive market for imported manufactured goods. This fact is central to Africa's current economic crisis. Only those countries which have substantial levels of resource endowment, coupled with small and highly skilled populations – for example, Botswana, Mauritius, South Africa ... and Equatorial Guinea – demonstrate any significant capacity to maintain long-term economic growth. (UN-HABITAT 2003)

African economies are largely service economies, with services contributing on average 69 per cent of GDP in 1998 and agricultural value added on average only 19 per cent (UNDP 2000). Therefore urban African economies are also service cities. However, the absence of a dynamic diversified economy has hindered the growth of higher-level business and financial services. The service sector is largely public administration, social services, personal services, trade and repair services. Cutbacks in public administration and sometimes social services have been part of structural adjustment, leading to declining employment. Some but not all of those made redundant have established themselves in private business. More dynamism might be expected in small enterprises with the entry of more qualified people, but little information on this seems to be available, and of course not all public sector employees are likely to succeed as entrepreneurs. In the early years after independence in many countries, especially in eastern and central Africa, the development of wholesale and retail trade was hindered by the large role taken on by the public sector, in some cases to break the hold of Asian businesspeople over this sector. Unlike in West Africa and countries such as Kenya, trade has only become a dynamic sector for private enterprise in more recent years in some of these countries. Repair services are common but have been hindered and often still are by the lack of available parts and equipment and the limited skills base.

'In the 1990s, African formal labour markets have been absorbing less than 25 per cent of the newcomers' (Hall and Pfeiffer 2000). Returns to education have fallen as a result. Private sector stagnation and capital flight have been exacerbated by public sector retrenchments, resulting in increased unemployment and reduced security in the formal labor market. One estimate of unemployment in the mid-1990s put it at about 20 per cent, 60–75 per cent of whom are young, even though young people only account for a third of the labor force. While unemployment is a concept of doubtful value in countries with few social safety nets, the predominance of young people amongst the unemployed is neither restricted to Africa nor new. Insecurity in the formal labor market has blurred any distinctions between the formal and informal sectors as large numbers have moved into casual or self-employment, either full-time or part-time to support falling real wages.

It has been estimated that at least half of Africa's urban workforce is engaged in so-called 'informal sector activities' (48 per cent in 1995, varying from 63 per cent in West Africa to 50 per cent in East Africa, 35 per cent in Middle Africa and 10 per cent in Southern Africa) (Hall and Pfeiffer 2000, based on ILO employment figures). These figures are borne out by those for particular cities (e.g., Bamako 83 per cent in 1990, Lagos 69 per cent in 1990, 65 per cent in Abidjan, 60 per cent in Ouagoudougou, 51 per cent in Niamey, 38 per cent in N'Djamena, 15 per cent in Khartoum) (Hall and Pfeiffer 2000; World Bank 2000b; Montgomery *et al.* 2004). The figures may well be underestimates, because many in full-time wage employment are also engaged in informal sector activity to make ends meet, and the estimate for Southern Africa appears surprisingly low. UN-HABITAT (2003) quotes a figure produced by *The Economist*, which claimed that informal sector activities typically added between 20 per cent and 30 per cent to African GDP in 1993.

Trade enterprises account for between half and 70 per cent of all informal sector enterprises (Rogerson 1997). At least half the enterprises are single person businesses, and what growth occurs takes place by replication and diversification rather than individual business expansion, in order to spread risk. Women have, with few exceptions, been disadvantaged in formal labor markets and as a result are over-represented in the informal sector, in the smaller and less profitable enterprises in particular, due to their lower levels of skill and even more limited access to capital than men. The retrenchment of formal sector employees is likely to have had both positive and negative impacts on the informal sector, bringing new capital, ideas and skills, but also increasing competition for those already in the sector. Liberalization is likely to have had both positive and negative effects, that is, activities that were formerly reserved for public sector enterprises are now open to private businesses and import restrictions have been lifted, but also the increased economic instability and decreased demand resulting from of high inflation, rising interest rates and falling real wages may have harmed micro-enterprises just as much as larger firms. Similarly, the informalization of the formal sector may have had both positive and negative outcomes – it may provide opportunities for subcontracting or commission selling (e.g., Rogerson 1997) but also may reduce the costs of formal enterprises relative to those of informal businesses.

When African urban economies are largely formal, for example, South Africa, Zambia to the 1980s, Zimbabwe to the 1990s, a distinction between the formal and informal sectors appeared to make sense, since demand for informal sector products and services tended to be generated almost entirely from formal sector wages. Today, however, the economies of many African cities are predominantly informal. It is unclear either that the remaining formal sector can generate sufficient demand to explain the continued existence and growth of informal sector enterprises or whether the informal sector contains potential for accumulation and economic growth or is merely a subsistence sector of last resort. Our conceptual tools for understanding the structure and dynamics of African urban economies are totally inadequate, not least because the information base is also totally inadequate. Many statistical series (e.g., labor force or employment and earnings surveys) either cover only part of the urban economy (usually the formal part), or were suspended temporarily or permanently in the 1980s. Most other data collection exercises have also concentrated on only parts of the urban economy – with the exception of an ILO study of urban labour markets in some West African cities in the 1990s (Lachaud 1994) – while, with the exception of a few enterprise clusters (e.g. Brautigam 1997), studies of enterprises of all scales within particular sub-sectors or local economies are relatively rare.

Thus, 'African cities are marginalized in the new global economy. African cities are growing despite poor macroeconomic performance and without the benefit of significant FDI in their economies' (Montgomery *et al.* 2004). In part, the poor economic performance of urban economies is linked to limited national economic growth, but it can also be attributed to failure to provide a conducive urban business environment. Economic development policies and investment have traditionally been regarded as national responsibilities. Local economic development is not a traditional local authority responsibility, so municipalities have neither the legislative base nor the expertise to do it. As a result, governance failures and infrastructure deficiencies mean that the environment provided by African urban areas is not conducive to efficient business operation, leaving them behind in the competition for private investment. To tackle this issue will require not only better analytical tools, which recognize that the so-called formal and informal sectors are related to each other by ties of supply and demand and by people engaged in both types of activity, but also improvements to governance and service delivery.

Governance Failures

Local authorities, whether urban or rural, are traditionally conservative, cautious, non-developmental, under-funded, bogged down with inappropriate regulations and unaccountable. Governance arrangements have seesawed between greater and lesser degrees of democracy and decentralization in most Sub-Saharan African countries. In the vast majority of countries, post-independence attempts to establish democratic politics foundered – when the management of multi-party competition proved problematic, elected governments were displaced by military coups or replaced by one-party rule. Although local government arrangements had been established with greater or lesser degrees of autonomy and democracy depending on the colonial power concerned, in the years leading up to independence, many countries (especially in Francophone Africa) inherited centralized systems and others centralized in the years after independence in order to forge national unity, achieve economic aims and control over the national space, and also to economize, given the limited human resources available to run government. After years of authoritarian centralized rule, the 1990s were marked by a new wave of democratization and decentralization, driven in part by internal pressures and in part by external influence. It is still too early to judge the effectiveness of the new political and administrative systems. Certainly, many of the problems that have characterized local politics and government since the 1960s still persist.

The first problem arises from a general ambivalence towards multi-party democracy and in particular central government ambivalence towards elected local governments. A legacy of authoritarian politics and state-led approaches to development seems to result in a limited appreciation of the need to secure public support for government policies, particularly if organized civil society only played a limited role in the transition to democracy. There are considerable differences between political systems such as those in Ghana, Zambia or Kenya, where organized civil society was frowned upon during one-party rule, is not directly represented in legislative bodies, and is only consulted when under pressure rather than accepted as a human right, and those of a country like South Africa, where organized civil society played a key role in overthrowing apartheid and has won a much more central role in central and local politics as a result (Rakodi 2004).

Central politicians fear the potential for local politics to provide a base for opposition challenges to local government. Indeed, it is common in multi-party systems – not only in Africa – for the cities to come under opposition control. One of the results seems to be a democratic deficit. In Kenya, for example, multi-party elections were restored in 1992 in response to domestic and external pressure, but for a decade, the electoral rules were abused. Opposition parties were sometimes refused permits to operate and were denied access to the media and there was violence, harassment and gerrymandering. Wards in some cities may have a larger average number of voters than others elsewhere in the country (e.g., in Mombasa, each councillor represents an average of nearly 30,000 residents compared to about 15,000 in Johannesburg or Kumasi). Boundaries of local authorities or constituencies may not be redrawn in line with increases in urban population. In addition, central governments may reserve for itself the right to nominate the chief executive or additional councilors to local authorities. In Ghana, for example, the President appoints municipal chief executives. From 1995–2001, Kumasi was governed in an autocratic, idiosyncratic and ineffective manner by an appointed mayor who was also a member of the local Asante royal family, as well as a supporter and financier of the ruling party and personal friend of the President. Presidential support, control over resources and the deference due to a member of the traditional elite enabled the Metropolitan Chief Executive (MCE) to command majority support in the elected Assembly or sideline it when necessary, despite widespread criticisms of his management and allegations of corruption. In Mombasa, the provision for seven nominated in addition to 24 elected councilors enabled the national government to secure a clear majority in the council (Rakodi 2004).

Related to the above is the reluctance of the central government to give local authorities autonomy. To some extent, this is related to the poor quality of local administration and the limited accountability of local government either to its own electorate or to central government. Central ministries are often unwilling to cede their functions, fearing loss of control over programs, loss of staff and reduced influence. In Kumasi, for example, central ministries are supposed to cede their functions to municipalities under the decentralization program. However, they have been reluctant to do so.

Central government also fears loss of control over revenue raising and expenditure. Not only do central governments, themselves chronically short of finance, keep control over the most lucrative and buoyant taxes, they give little attention to developing well-designed local government funding systems and are also reluctant to let municipalities develop their own revenue bases. Central government approval is generally required for local authorities to borrow, develop a new tax, update their valuation rolls, change the property tax rate, or increase user charges for services. Although large cities could generally raise a larger proportion of the resources they need from their own revenue, given the scope to do so, they are unlikely to be completely self-sufficient and smaller cities and towns will almost certainly need some central government funding. In these circumstances, the central government is justified in auditing local government spending and providing policy guidance to ensure that its funds are spent in line with national policy guidelines. However, arbitrary and slow processes of approval for local authority budgets and inefficient auditing encourage bad budgetary practice amongst municipalities (Devas 2004). In addition, unpredictable volumes and timing of central-local fiscal transfers hamper municipal attempts to plan ahead and address local priorities. Attempts to address these issues (e.g., in Uganda, Kenya, Tanzania or South Africa) are limited and too recent to be fairly evaluated.

The lack of autonomy, resources and capacity means that local politics is not particularly attractive, compared to national politics. It is often asserted that local politics, as a result, attract candidates and politicians of a lower caliber, with lower levels of education and expertise than those involved in national politics. As a result, local decision-making is of poor quality and the management of local authorities fails to improve. Some may see local politics as a stepping-stone to national politics and lack local commitment, some have a limited understanding of the role and functions of local government, others may use it primarily as a patronage resource. For example, in Mombasa, the number of unskilled employees of the council tended to increase prior to elections in the 1990s, despite the lack of resources or supervisory personnel to make good use of the additional personnel. In Kumasi in 1994, the Kumasi Metropolitan Assembly (KMA) decided to privatize the management of public toilets, some 240 of which provide sanitary facilities to low-income residents, in order to improve their management. This worked reasonably well for a while. However, in 1997, the MCE decided to allocate the contracts to those members of the KMA who supported him. Not only did charges go up but the operation and maintenance of the toilets also deteriorated, and required payments were not remitted to the KMA by operators (Korboe *et al.* 1999; King *et al.* 2001).

In the vast majority of African countries, parties in multi-party systems compete on bases other than a policy platform. It is extremely rare for parties putting forward candidates in local (or even national) elections to have a detailed manifesto, although this does not restrain candidates from making ambitious promises. Of course, external policy conditionality means that parties have very little choice over policies, which are largely dictated by the international financial institutions. Often elections are fought ostensibly to oust a party that has been ineffective and corrupt in office. Because the parties do not have distinct ideologies or policies, they tend to command loyalty on the basis of regional, religious or ethnic identity instead. National political rivalries are played out at the local level, although occasionally local candidates in good standing are voted into office in the hope that local service delivery will improve. In Mombasa in 1997, for example, a local businessman and member of the ruling party (KANU) was nominated as a councilor and subsequently by the councilors as mayor, in the hope that he could revive the flagging economic fortunes of Kenya's second city. However, his tenure was brought to an abrupt end when he crossed swords over alleged land grabbing with another businessman with stronger connections to the ruling party in Nairobi (Rakodi *et al.* 2000).

Many local electoral systems are based on a first-past-the-post (FPTP) model, and have the strengths and weaknesses of such systems worldwide. Ward-based elections do seem to improve accountability by councilors to residents, but FPTP systems also marginalize minorities and encourage short-term political time horizons, as those elected seek to reward their supporters in time to secure re-election. In some local government systems, however, attempts have been made to overcome these disadvantages by mixing FPTP and proportional representation (e.g., in the new South African local government system) or by reserving seats for under-represented groups (e.g., women, youth and the disabled in Ugandan national and central government).

In the largest cities, elected metropolitan government can be too remote, even though it may be desirable for strategic planning and infrastructure investment, and for cross-subsidy between commercial and high income residential and low income residential areas. One alternative is for smaller elected municipalities to collaborate. In Abidjan, for example, the metropolitan area (population 3.5 million) was divided into 10 municipalities in 1980. Each elects five councilors and a mayor who, together with a metropolitan mayor, are also members of the supra-municipal body, Ville d'Abidjan. The 11 mayors form an executive committee. This system has not, however, solved all the problems of overlapping responsibilities, weak metropolitan leadership, reluctance to devolve powers and responsibilities, and lack of coordination between central and local government and between municipalities with very different levels of wealth (Attahi 1999). An alternative is to establish sub-city levels of government. However, city-level governments may see them as a threat, with the result that they also have unfunded mandates, little standing in the eyes of residents, and limited effectiveness (e.g., in Kumasi). Also, local administration may be an instrument of command and control on the part of central government rather than a means for local representation and decision-making. Its local role then depends on the nature of national politics, often concentrating on securing spoils and dispensing patronage. For example, the local administration in Kenya is essentially an apparatus to deliver votes and maintain security. All except the lowest level (i.e., village elders) are appointed by central government, and the local chief and elders may or may not have legitimacy in the eyes of local residents.

Reliance on civil society to increase the responsiveness and accountability of local government is limited in the African context. Non-governmental organizations (NGOs) were discouraged and even actively repressed in many military and one-party regimes. Although the political space for their activities increased with democratization, lobbying organizations are still weakly developed and those that are active often focus their attentions on national policy. There is an imbalance between small struggling indigenous welfare or developmental NGOs and well-funded large Northern NGOs. The ability of civil society to ensure that the needs of needy groups are addressed is reduced by poor quality and non-transparent decision-making. Local authorities are often engaged in crisis management and strategic decision-making is limited (except, to some extent, in those cities supported by the Urban Management Programme and Cities Alliance to prepare city development strategies). Plans are out of date, decisions made without prior consultation and there are often tensions between elected councilors and local civil society organizations. Accounts, which are years out of date, and budgets, which are essentially fictional (amongst other reasons because of poor central-local fiscal transfer arrangements), make the exercise of citizen voice difficult, notwithstanding some recent NGO efforts to promote participatory and gender aware budgeting (e.g., FOWODE in Uganda or IDASA in South Africa) (One World Action 2002).

Community-based organization as a means through which citizens can exercise voice and development actions can be organized is also underdeveloped in many African towns and cities. In countries where indirect rule was employed during colonial times, traditional authorities have survived. With important functions related to administering land held by indigenous groups and families, these authorities and their constituent social groups (extended families, clans, etc.) continue to play an important role in urban development, for example, in many west and southern African cities (e.g., in Nigeria, Ghana, Senegal, Côte d'Ivoire, Botswana, Lesotho). Thus in the Côte d'Ivoire, the local chief combines traditional authority with administrative and organizational functions and is usually embedded in the long dominant structures of the single party (Berg-Schlosser 2003). Often the legitimacy of traditional authorities in the eyes of residents is stronger than that of government, but they may have an ambivalent relationship with elected representatives. In many one-party states, community-based organization was taken over by the party as a means of developing a grassroots organization for mobilization, control and, to a limited extent, participation in some aspects of local decision-making. Sometimes this form of community-based organization (CBO) had considerable importance and legitimacy in land management and local service delivery (e.g., Mozambique, Tanzania, Zambia), for example, during many externally funded squatter upgrading programs in the 1970s and 1980s. However, the link between CBO and the party was broken with the restoration of multi-party democracy. Sometimes, it appears to have developed considerable legitimacy in the eyes of local residents and to retain useful functions (e.g., in Dar-es-Salaam). Elsewhere, the notion that one party should control a residential area and will be the CBO in that area has persisted, even though the party in control may change following elections (e.g., in Lusaka).

Overall, similar to national governance, local governance in African towns and cities continues to be weak politically, financially and administratively. The vicious circle of pork-barrel politics, limited accountability, unfunded mandates and lack of administrative capacity reinforces the ambivalence of central government and local electorates alike to local government. Democracy, accountability and the provision of services as of right seem to be stronger where civil society organizations played an important role in the struggle for democracy (e.g., in South Africa). Elsewhere, local government still manages to deliver some services, undertake some developmental and regulatory functions and raise some revenue despite its chronic

shortage of resources, and in some places, both elected local government and civil
society have considerable vitality.

Infrastructure and Services

Governance failures and structural adjustment policies have contributed to the
failure of service provision to keep pace with urban growth and the declining
operational efficiency of installed infrastructure. Overall, in African urban areas,
44 per cent of residents lack on-plot provision of all three basic services (i.e., water,
sewerage, electricity). The situation varies greatly between cities, but overall is
worse in smaller towns and cities, and much worse for poor people than the non-
poor.

Table 3.2   Access to water, sanitation and electricity in urban areas

City size (inhabitants)	Piped/well water on premises*	Water in neighborhood	Toilet Flush	Pit	Electricity	Lacks three basic services**
Less than 100,000	35.4	n.a.	18.0	65.7	33.8	50.1
100,000–500,000	45.2	n.a.	20.6	70.6	46.7	41.1
500,000–1 million	42.3	n.a.	25.4	72.6	52.0	34.5
1–5 million*	55.5	n.a.	63.5	65.5	21.7	
Total urban	40.8	50.1	21.7	66.8	41.5	43.9
Urban poor***	26.9	61.6	13.0	65.9	19.7	62.9
Urban non-poor	47.6	45.8	27.4	67.2	52.2	34.2

Notes: * Excludes Lagos; ** piped water, flush toilet, electricity; *** Bottom quartile of an
index of household welfare based on consumer durables and housing quality, derived using
principal components analysis.
Source: Montgomery *et al.* 2004. The figures are based on between 20 and 30 Demographic
and Health Surveys (DHS).

Fewer than half of urban residents in African towns and cities have on-plot water
supply, but for many of these the supply is well water rather than piped water. There
has been little improvement in recent decades and the situation is worse in smaller
towns and cities and for poor residents (see Table 3.2). There is nothing unusual in
deficiencies of urban water supply – this is typical of rapidly growing cities
worldwide. However, even in cities where the main piped supply has broken down
or failed to expand to serve new areas and local income residents, alternative
arrangements have evolved, enabling cities to survive despite the breakdown of
large-scale reticulated supply systems. Individual strategies have included shallow
wells or boreholes – the latter adopted by high-income residents, industries and
hotels but also sometimes by low-income communities, linked to a limited piped
network. Small-scale private sector operators have also evolved roles in water
distribution. These entrepreneurs (or community groups) may operate kiosks where

they sell water from the piped supply or become mobile vendors. Although vital to the survival of urban populations, such systems have disadvantages, including overpumping leading to the lowering of the water table, and the water they supply to the poor may be poor quality and costly, especially if the volume of water in the piped supply or the number of outlets are limited. Large-scale private sector participation as a means of overcoming the perceived deficiencies of public water and sewerage utilities has a relatively long history in Francophone African cities, but has been tried only recently and on a limited scale elsewhere. There have also been experiments with corporatization of publicly owned utilities – or at the very least management autonomy. There are examples of cities with both privately and publicly managed water supplies that have achieved a relatively good performance (e.g., in Côte d'Ivoire, Zimbabwe), although the jury is still out on most recent experiments.

Sewerage systems serve only limited areas of most cities (see Table 3.2), are costly, and have ceased to function reliably in many places (because of interruptions to water supply, lack of imported chemicals for sewage treatment, etc.) or have broken down. However, waterborne sewage is arguably an inappropriately costly technology for low-density cities in any case. Because coverage has been so limited, most residents depend on septic tanks or pit latrines. The former does depend, of course, on a piped water supply and a tank evacuation service. With the opening up of the latter to private operators, the availability of tank evacuation services has improved, although controls over where the tankers discharge their contents are often weak. Septic tanks can also provide a means of sanitation in low-income areas, where sometimes shared toilets work and sometimes they do not. Pit latrine sanitation with improved technology is generally adequate, however, provided piped water is available. Nevertheless, evacuation services suitable for emptying pit latrines in very high-density areas with limited access ways are often not available and requirements for provision by landlords are not enforced (e.g., in Kibera in Nairobi).

Electricity is generally not a local authority responsibility. Generally supplies are deficient and power outages frequent. However sometimes it has been made more widely available in the informal settlements where most poor people live than other services (although Table 3.2 shows that there are still gross inequalities between poor and non-poor residents), partly because it is an above ground service and so is technically easier to install, partly because it does not seem to imply de facto recognition of 'illegal' and 'unplanned' settlements in the same way as water, and also because cost recovery from users is relatively easy. Telephones, on the other hand, never did reach the majority of urban residents, especially in low-income areas, and landline technology has now been overtaken by mobile phones, which have been made available by small businesses in most areas where reception is possible.

Roads, never provided to many informal settlements, have deteriorated into potholed highways and rutted tracks, under the combined impact of lack of finance and poor maintenance, until an external donor can be prevailed upon to grant or lend funds for rehabilitation and reconstruction. Finally, solid waste management has also been deficient, because of limited local government capacity and failure to recognize and improve access to informal settlements. According to one estimate, less than a third of residents in urban areas in Sub-Saharan Africa have a formal solid waste collection service (UN-HABITAT 2003). Medium-scale private sector participation has improved services in some commercial and high-income areas, but often service improvements have been constrained by the lack of local government capacity to draw up contracts and monitor performance. The scope for community contracting is currently being explored on a small scale (e.g., in Dar-es-Salaam). The deficiencies in environmental health services have adverse impacts both on the health of urban populations (especially the residents of informal settlements) and on the wider environment, including groundwater, rivers and streams, and coastal waters.

Often the infrastructure and service deficiencies in African towns and cities occur because of underfunding and central government controls that prevent municipalities addressing the problems. However, many municipalities also lack the capacity to undertake the necessary planning, investment, operation and maintenance, either directly or through contracting arrangements of different sorts. In addition, the rigidity of professionals such as engineers and planners in insisting on adherence to relatively high standards, the rationale for which is often not clear and which are costly and unrealistic, is a major contributory factor. The deficiencies in planned provision and emergence of alternative arrangements in some cases are closely linked to problems of land delivery, which will be discussed in the next section.

Policing is inadequate, unaccountable, corrupt and often violent in many African towns and cities. Heightened insecurity affects where people can go and what they can do – the more everyday life is withdrawn from public spaces, the more dangerous they become. However, local solutions such as neighborhood vigilante groups or police crackdowns can contribute more to intensifying insecurity and arbitrary justice than reassuring frightened populations or reducing crime, since the resentment generated precipitates further criminality (Simone 2002).

> In Kinshasa, the forces of law and order are themselves one of the main agents of violence … Violence, crime, abuse of human rights, and extra-legal actions on the part of the police and military forces are increasingly widespread. The majority of Kinois involved in any type of conflict do not call on the justice system to resolve their problems. The widespread feeling is that the system is extremely biased and does not offer poor people any possibility of justice. (Nlandu 2002)

However, improvements are possible, as illustrated by the public revolt, immediately following Kenya's 2001 change of government, against the extraction of bribes by police at roadblocks, which appears to have halted the practice overnight.

Social services such as health care and education are also deficient in most towns and cities, not least because of the impacts of conditionality with respect to budget deficits. Overall just less than 80 per cent of urban children aged 9 to 10 is attending school, but only 57 per cent of those aged 15 to 16. Of the latter group, 43 per cent in poor households compared to 62 per cent in non-poor households are enrolled in school and the proportion of boys from both groups exceeds the proportion of girls (Montgomery *et al.* 2004). The proportion of women whose recent births had all been attended by a qualified health professional was 32 per cent for poor women and 41 per cent for non-poor women, a general figure that is supported by recent figures for Nairobi which show that while only 54 per cent of women in Nairobi slums had been attended by health personnel, 76 per cent of women in Nairobi as a whole had been able to access health services for this purpose (Montgomery *et al.* 2004; APHRC 2002). Social sector spending has not always declined, but even where it has not done so, increased cost recovery – both official and unofficial – has increased the cost of services to users. The result is poor quality services, sometimes at a relatively high cost to users. Private alternatives have proliferated, resulting in more widespread availability, but many of these services are high cost and, in the absence of regulation, their quality is extremely patchy. The result has been increased inequality, as illustrated by various health indicators. For example, while average urban mortality is generally lower than rural mortality, this does not apply to mortality rates in slums or smaller cities. Infant mortality rates are 81 out of 1000 live births in urban areas overall but 89 out of 1000 for the urban poor compared to 74 out of 1000 for the non-poor (Montgomery *et al.* 2004).

Land Delivery and Administration

The formal tenure and administration systems inherited from colonial administrations and thought to be more appropriate for urban development than indigenous systems have proved unable to cope with rapid urban growth. As a result, between 50 and 70 per cent of all urban land is delivered in the vast majority of Sub-Saharan African cities and towns through informal systems. These include a variety of different channels. The presence and relative importance of alternative channels depends on the following factors:

- Pre-colonial sociopolitical structures, in particular land allocation and administration arrangements which prevailed;
- The choice made by colonial powers between direct and indirect rule, which was influenced by the pre-colonial sociopolitical structure, as well as the colonial powers' conception of how to rule their empires, the need and scope for direct settlement, etc.;
- The location of the urban center, in particular whether it is surrounded by an area expropriated for commercial farming and therefore subject to imported tenure and administrative arrangements, or by an area of indigenous tenure such as a native reserve;
- Post-colonial land reforms, particularly whether land was nationalized in the 1970s, with subdivision and allocation reserved for the state, and whether attempts were made to develop alternatives to individual plot surveys and titles.

Recent research in six cities found a variety of channels of informal land delivery, varying in their importance, strengths and weaknesses and relationship to the formal land administration system.

Non-commercial and semi-commercial channels:

- Squatting: In the years leading up to and after independence in many countries squatting was widespread. It occurred where there were abandoned farms, where private ownership rights were not strongly entrenched, where it was politically tolerated, where there were extensive areas of land in public ownership, where the population was growing rapidly and where there was a large shortfall between the supply of public housing and demand (e.g., Zambia, but not Kenya or Zimbabwe). Today, the scope for squatting is extremely limited, not least because most publicly owned land has either been developed or occupied. The areas settled earlier have often been regularized, and markets in developed property and rental housing have emerged. It occurs in a few cities (e.g., Kampala) on a small scale on the remaining publicly owned land, or on sites considered unsuitable for residential use – either because they are topographically unsuitable, for example, liable to flooding, or because they are on land reserved for other uses, for example, under power lines.
- Family or group land: It is possible for family or group members (mainly men) to obtain land free from family or group reserves in those cities where land remains under the control of indigenous land rights holders. In some instances, sites in the central parts of such cities (e.g., Enugu) have been fully developed and this is no longer possible for some groups. The desire to retain sufficient family land for succeeding generations has influenced the plot size preferred by indigenous subdividers (e.g., in Maseru) and the length of occupancy they are prepared to grant others (e.g., on the outskirts of Enugu).
- Party sanctioned land allocation (e.g., Lusaka).
- Grabbing and/or illegal subdivision of publicly owned land (e.g., Kenya until the recent change of government).

Commercial channels:

- Sale of family or group land is the main channel of land delivery in those cities surrounded by land over which indigenous groups are rights holders (e.g., Maseru, Enugu, Kampala). Arrangements for subdivision and disposal of such land have sometimes received state sanction or been subject to state interference (e.g., in Kumasi or Kampala) and in other cases are under the control of individuals (e.g., arable field owners in Maseru), local groups (e.g., families in Enugu) and/or traditional authorities (e.g., chiefs in Maseru or Dakar; see Abdoul 2002; the Buganda Kingdom in Kampala).
- Illegal subdivision of privately owned land, usually because the processes of subdivision and tenure registration are slow and cumbersome (e.g., land-buying companies in Eldoret, Kenya).

These channels produce a large volume of land for urban development. While in the early days after independence, widespread squatting and subdivision of land by indigenous right holders provided access to land for recent migrants and the poor, this is generally no longer the case. Most land in most cities is now delivered through commercial channels, and only available to those with means – although they are less discriminatory against women than indigenous systems. The processes involved are regulated by social institutions; some are based on traditional social organizations and processes and others are modern inventions. Some have been instituted by a recognized social organization, such as a traditional authority or a political party, while others have evolved through multiple social interactions between informal social groups (families, local leaders). Particular arrangements which have evolved include informal written agreements between sellers and purchasers; the practice of witnessing by family members, local residents, local level government officials and/or party functionaries; and the institution of written records (e.g., by extended family groups in Enugu, the lowest levels of local government in Kampala). Often the institutions that regulate transactions in land work quite well, but they also have limits. Sometimes these emerge when densities increase and earlier ways of regulating land use and resolving disputes break down (e.g., in inner city Dar-es-Salaam; see Kombe and Kreibich 2000), although this is not always the case. Elsewhere, insecurity may increase when multiple sales of the same plot occur or earlier transactions are challenged by other family members. The reaction is often to develop a link to the formal state land administration system. For example, in Enugu buyers seek to register their ownership as a title. Sometimes, this takes the form of the formal courts recognizing informal documents such as written agreements between purchasers and sellers. Sometimes government regularizes an informal settlement, either *de facto* (by providing services) or *de jure*.

Many features of these informal delivery systems, the social institutions giving transactions legitimacy, the mechanisms for resolving disputes and the mutually beneficial links they have developed with the formal system have demonstrated considerably more promise as ways of delivering large quantities of land for urban development than the formal systems themselves. Obstacles to learning from what works and addressing the weaknesses in a flexible way include the rigid thinking of professionals (i.e., planners, surveyors and engineers), particularly their adherence to unrealistic standards and procedures. Sometimes, political decision-making, generally at the national level, is also an obstacle. A preoccupation with rural land may have prevented attention being paid to sorting out arrangements for urban land (e.g., in Uganda) or political struggles over the role of traditional authorities in government may raise the political temperature over all issues that are of concern to the traditional authorities (e.g., Maseru). However, there are examples of innovative approaches to improving land delivery that learnt from aspects of the informal systems, for example, the use of occupancy licenses in Zambian informal settlements to provide security, negotiating with land buying companies to reserve sites for social facilities and access in Eldoret.

Poverty and Social Breakdown

In a few extreme cases (e.g., Luanda), urbanization has been associated with extreme and widespread poverty and social breakdown. However, the discussions of land delivery and governance arrangements above have illustrated that, although poverty has often become more widespread with economic decline and structural adjustment, a more nuanced picture of urban society is needed.

Family and kinship networks are generally regarded as the basic support networks for African people, responsible for socializing children, caring for dependents and providing access to opportunities and support networks. There is widespread concern that these are breaking down under the impact of successive shocks and prolonged economic stress on the one hand and urbanization on the other. Views vary from alarmist to overly optimistic, often based on anecdotal rather than systematic evidence. Apparently increasing problems of street children, youth unemployment, drugs and crime give rise to concern. Also, increasing numbers of AIDS orphans clearly place a burden on grandparents and other members of extended families. The expectation that urban families will support recent migrants while they find work and housing has been tricky since the 1960s. Declining real wages and increased redundancies since the 1980s have exacerbated the strains on urban families, but most appear to have found ways of managing the reciprocal obligations involved. Nevertheless, for many, family and kin still appear to be the key social units and support networks, straddling rural and urban areas and performing both social and economic functions. Often, they are linked to other forms of social organization, which are also important to fulfilling social obligations (such as burial societies or home town associations; see Adetula 2002) or to household livelihood strategies (such as Rotating Savings and Credit Association (ROSCAs)). Other people, including some whose families have been urban residents for several generations now, have, at least in part, substituted other social networks (e.g., professional or trade associations, workplace-based ROSCAs).

Linked to the continued importance of kinship are traditional authority structures, which continue to be important in countries where they were not deliberately destroyed by the colonial authorities. They are generally socially and culturally important, even in urban areas, and command higher levels of legitimacy and trust than many government or political organizations (Abdoul 2002). However, they are also conservative, hierarchical and patriarchal, maintaining and reinforcing deference and women's subordination.

Generally, local organizations seem to have been more commonly initiated by outside agencies than evolved locally. They have been established by the ruling political party (e.g., in Tanzania, Zambia and Mozambique), by the central government (e.g., in Uganda) or by NGOs (e.g., in Zambia since democratization). They vary in their form, legitimacy, capacity, stability and relationship to the political and governmental system. Organizations in high-income areas or of professionals or larger businesses tend to have more resources, including political access, and to have greater influence on decision-making, as illustrated by the Sandton Ratepayers Association boycott which worsened Johannesburg's financial crisis a few years ago. Many CBOs emerge to resist an external threat (e.g., eviction) or respond to an external initiative (e.g., regularization and upgrading) and may not persist (or may go dormant) after some time, especially if they cannot obtain access to ongoing resources. The relationships between social organizations and the political system may be conflictual, consultative or aimed at making claims and securing rights.

Other social structures that are important to residents may not play a role in local governance, for example, religious organizations, although they may also be involved in welfare and delivering services (e.g., health care services, schools), often as an alternative to state-provided services. Moreover, in many towns and cities new social dynamics have emerged around creative activities, including drama, music and art, which have evolved new and distinctive forms and sometimes provided channels for political protest, even while they also give expression to urban fears, pressures and the anxieties involved in much day-to-day urban life (De Boeck 2002; Enwezor *et al.* 2002). Often young people are excluded from any meaningful participation in decision-making. They have come to rely on the street, music, dance, dress and language to express themselves and demonstrate their disaffection (Simone 2002).

Many of the organizations that are important to urban residents provide social support, but aspects of the relationships involved can be repressive, as has been mentioned above with respect to families, kinship and traditional structures, and they can also be exclusive, reproducing the gender, ethnic and class cleavages present in wider society and damaging social cohesion (Tostensen *et al.* 2001). In addition, if they are forms of social organization whose members are mostly poor, the bonding social capital they provide is unlikely to provide opportunities for people to improve their situations, unlike forms of organization that incorporate people from different socioeconomic categories and can provide bridging social capital. The latter include kinship networks in some circumstances, but also include patron-client relationships with politicians or bureaucrats, which may provide access to sufficient resources to hold back political movements advocating the provision of services as of right. In addition, some forms of social organization (e.g., criminal gangs) can be disruptive, resulting in increased levels of crime and violence, and increasing people's vulnerability.

Conclusion

What is the future for African cities and towns? Are they socially, politically, administratively, economically and environmentally sustainable? Or are they in a situation of crisis, which can only worsen in the future? We can only speculate. This brief and sketchy overview has tried to demonstrate that African urban centers are indeed facing serious challenges, and some of them are in crisis, especially those affected by large-scale civil conflict and government collapse. Inevitably, their fortunes are bound up with the fortunes of the countries in which they are located, and the prospects for some countries involved in war, with minimal economic resources or with authoritarian and destructive regimes are extremely poor. For many countries, the future is highly uncertain, given their adverse incorporation and marginalization in the global economy, the limited decision-making autonomy of highly indebted countries, and the difficulties incipient democracies face attaining stability and peaceful transfers of government when economic resources are extremely limited. In others, however, there are promising developments with respect to consolidation of democracy, adaptation of democratic forms to local circumstances, economic stability or recovery, public sector reforms including decentralization, and the emergence of a stronger and more engaged civil society. In towns and cities, in addition to governance failures, severe infrastructure deficiencies and inadequate land administration systems, there are manifestations of resilience and creativity, especially in family and kinship structures, cultural and associational life, and indigenous social organizations and institutions. In addition, there is considerable potential for learning from and building on evolving informal systems and the interfaces they are evolving with formal systems.

PART II
NEW SCALES IN
SPACE AND TIME

Chapter 4

The Global City:
Strategic Site, New Frontier

Saskia Sassen

The master images in the currently dominant account of economic globalization emphasize hypermobility, global communications, and the neutralization of place and distance. There is a tendency to take the existence of a global economic system as a given, a function of the power of transnational corporations and global communications.

But the capabilities for global operation, coordination and control contained in the new information technologies and in the power of transnational corporations need to be produced. By focusing on the production of these capabilities, we add a neglected dimension to the familiar issue of the power of large corporations and the new technologies. The emphasis shifts to the practices that constitute what we call 'economic globalization' and 'global control': the work of producing and reproducing, the organization and management of a global production system and a global marketplace for finance, both under conditions of economic concentration.

A focus on practices draws the categories of place and production process into the analysis of economic globalization. These are two categories easily overlooked in accounts centered on the hypermobility of capital and the power of transnationals. Developing categories such as place and production process does not negate the centrality of hypermobility and power. Rather, they bring to the fore the fact that many of the resources necessary for global economic activities are not hypermobile and are, indeed, deeply embedded in place, notably often global cities and export processing zones.

Why is it important to recover place and production in analyses of the global economy, particularly as these are constituted in major cities? Because they allow us to see the multiplicity of economies and work cultures in which the global information economy is embedded. They also allow us to recover the concrete, localized processes through which globalization exists and to argue that much of the multiculturalism in large cities is as much a part of globalization as is international finance. Finally, focusing on cities allows us to specify a geography of strategic places at the global scale, places bound to each other by the dynamics of economic globalization. I refer to this as a 'new geography of centrality', and one of the questions it engenders is whether this new transnational geography also is the space for new transnational politics. Insofar as my economic analysis of the global city recovers the broad array of jobs and work cultures that are part of the global economy though typically not marked as such, it allows me to examine the

73

possibility of a new politics of traditionally disadvantaged actors operating in this transnational economic geography. This is a politics that lies at the intersection of economic participation in the global economy and the politics of the disadvantaged, and in that sense would add an economic dimension, specifically through those who hold the other jobs in the global economy – whether factory workers in export processing zones in Asia, garment sweatshop workers in Los Angeles or janitors on Wall Street.

These are the subjects addressed in this chapter. The first section examines the role of production and place in analyses of the global economy. The second section posits the formation of new geographies of centrality and marginality constituted by these processes of globalization. The third section discusses some of the elements that suggest the formation of a new socio-spatial order in global cities. The fourth section discusses some of the localizations of the global by focusing particularly on immigrant women in global cities. In the final section, I discuss the global city as a nexus where these various trends come together and produce new political alignments.

Place and Production in the Global Economy

Globalization can be deconstructed in terms of the strategic sites where global processes and the linkages that bind them materialize. Among these sites are export-processing zones, offshore banking centers and on a far more complex level, global cities. This produces a specific geography of globalization and underlines the extent to which it is not a planetary event encompassing the entire world.[1] It is, furthermore, a changing geography, one that has transformed over the last few centuries and over the last few decades.[2] Most recently, this changing geography has come to include electronic space.

The geography of globalization contains both a dynamic of dispersal and of centralization, a condition that is only now beginning to receive recognition.[3] The massive trends towards the spatial dispersal of economic activities at the metropolitan, national and global level that we associate with globalization have contributed to a demand for new forms of territorial centralization of top-level management and control operations. The spatial dispersal of economic activity made possible by telematics contributes to an expansion of central functions if this dispersal is to take place under the continuing concentration in control, ownership and profit appropriation that characterizes the current economic system.[4]

National and global markets as well as globally integrated organizations require central places where the work of globalization gets done.[5] Furthermore, information industries also require a vast physical infrastructure containing strategic nodes with a hyperconcentration of facilities; we need to distinguish between the capacity for global transmission and communication and the material conditions that make this possible. Finally, even the most advanced information industries have a production process that is at least partly bound to place because of the combination of resources it requires even when the outputs are hypermobile.

The vast new economic topography that is being implemented through electronic space is one moment, one fragment, of an even vaster economic chain that is in good part embedded in non-electronic spaces. There is no fully dematerialized firm or industry. Even the most advanced information industries, such as finance, are installed only partly in electronic space. And so are industries that produce digital products, such as software designers. The growing digitalization of economic activities has not eliminated the need for major international business and financial centers and all the material resources they concentrate, from state of the art telematics infrastructure to brain talent (Castells 1989; Graham and Marvin 1996; Sassen 1998: Chapter 9).[6]

In my research, I have conceptualized cities as production sites for the leading information industries of our time in order to recover the infrastructure of activities, firms and jobs that is necessary to run the advanced corporate economy, including its globalized sectors.[7] These industries are typically conceptualized in terms of the hypermobility of their outputs and the high levels of expertise of their professionals rather than in terms of the production process involved and the requisite infrastructure of facilities and non-expert jobs that are also part of these industries. A detailed analysis of service-based urban economies shows that there is a considerable articulation of firms, sectors, and workers who may appear as though they have little connection to an urban economy dominated by finance and specialized services but, in fact, fulfill a series of functions that are an integral part of that economy. They do so, however, under conditions of sharp social, earnings, and often racial/ethnic segmentation (Sassen 2001: Chapters 8 and 9).

In the day-to-day work of the leading services complex dominated by finance, a large share of the jobs involved is low paid and manual, many held by women and immigrants. Although these types of workers and jobs are never represented as part of the global economy, they are in fact part of the infrastructure of jobs involved in running and implementing the global economic system, including such an advanced form as international finance.[8] The top end of the corporate economy – the corporate towers that project engineering expertise, precision, 'techne' – is far easier to mark as necessary for an advanced economic system than are truckers and other industrial service workers, even though these are a necessary ingredient.[9]

We see here a dynamic of valorization at work that has sharply increased the distance between the devalorized and the valorized, indeed overvalorized, sectors of the economy.

For me as a political economist, addressing these issues has meant working in several systems of representation and constructing spaces of intersection. There are analytic moments when two systems of representation intersect. Such analytic moments are easily experienced as spaces of silence, of absence. One challenge is to see what happens in those spaces or what operations – analytic, of power, of meaning – take place there.

One version of these spaces of intersection is what I have called 'analytic borderlands' (Sassen 1998: Chapter 1). Why borderlands? Because they are spaces that are constituted in terms of discontinuities – discontinuities are then given a terrain rather than reduced to a dividing line. Much of my work on economic globalization and cities has focused on these discontinuities and has sought to reconstitute them analytically as borderlands rather than dividing lines. This produces a terrain within which these discontinuities can be reconstituted in terms of economic operations whose properties are not merely a function of the spaces on each side (i.e., a reduction to the condition of dividing line) but also, and most centrally, of the discontinuity itself, the argument being that discontinuities are an integral part, a component, of the economic system.

A New Geography of Centers and Margins

The ascendance of information industries and the growth of a global economy, both inextricably linked, have contributed to a new geography of centrality and marginality. This new geography partly reproduces existing inequalities but also is the outcome of a dynamic specific to current forms of economic growth. It assumes many forms and operates in many arenas, from the distribution of telecommunications facilities to the structure of the economy and of employment. Global cities accumulate immense concentrations of economic power while cities that were once major manufacturing centers suffer inordinate declines; the downtowns of cities and business centers in metropolitan areas receive massive investments in real estate and telecommunications while low-income urban and metropolitan areas are starved for resources; highly educated workers in the corporate sector see their incomes rise to unusually high levels while low- or medium-skilled workers see theirs sink. Financial services produce superprofits while industrial services barely survive.[10]

The most powerful of these new geographies of centrality at the global level binds the major international financial and business centers: New York, London, Tokyo, Paris, Frankfurt, Zurich, Amsterdam, Los Angeles, Sydney, Hong Kong, among others. But this geography now also includes cities such as Bangkok, Taipei, São Paulo and Mexico City (Sassen 2000, GAWC; Sassen 2002). The intensity of transactions among these cities, particularly through the financial markets, trade in services and investment has increased sharply, and so have the orders of magnitude involved (e.g., Noyelle and Dutka 1988; Knox and Taylor 1995).[11] At the same time, there has been a sharpening inequality in the concentration of strategic resources and activities in each of these cities compared to that of other cities in the same country.[12]

Alongside these new global and regional hierarchies of cities is a vast territory that has become increasingly peripheral and increasingly excluded from the major economic processes that are seen as fueling economic growth in the new global economy. Formerly important manufacturing centers and port cities have lost functions and are in decline, not only in the less developed countries but also in the most advanced economies. Similarly in the valuation of labor inputs: the overvalorization of specialized services and professional workers has marked many of the 'other' types of economic activities and workers as unnecessary or irrelevant to an advanced economy.

There are other forms of this segmented marking of what is and what is not an instance of the new global economy. For example, the mainstream account on globalization recognizes that there is an international professional class of workers and highly internationalized business environments due to the presence of foreign firms and personnel. What has not been recognized is the possibility that we are seeing an internationalized labor market for low-wage manual and service workers;

or that there is an internationalized business environment in many immigrant communities (e.g. Ehrenreich and Hochschild 2003). These processes continue to be couched in terms of immigration, a narrative rooted in an earlier historical period.

This signals that there are representations of the global or the transnational that have not been recognized as such or are contested. Among these is the question of immigration, as well as the multiplicity of work environments it contributes in large cities, often subsumed under the notion of the ethnic economy and the informal economy. Much of what we still narrate in the language of immigration and ethnicity I would argue is actually a series of processes having to do with (1) the globalization of economic activity, of cultural activity, of identity formation, and (2) the increasingly marked racialization of labor market segmentation so that the components of the production process in the advanced global information economy taking place in immigrant work environments are components not recognized as part of that global information economy. Immigration and ethnicity are constituted as otherness. Understanding them as a set of processes whereby global elements are localized, international labor markets are constituted and cultures from all over the world are de- and re-territorialized, puts them right there at the center along with the internationalization of capital as a fundamental aspect of globalization.[13]

How have these new processes of valorization and devalorization and the inequalities they produce come about? This is the subject addressed in the next section.

Elements of a New Socio-spatial Order

The implantation of global processes and markets in major cities has meant that the internationalized sector of the urban economy has expanded sharply and has imposed a new set of criteria for valuing or pricing various economic activities and outcomes. This has had devastating effects on large sectors of the urban economy. It is not simply a quantitative transformation; we see here the elements for a new economic regime.

These tendencies towards polarization assume distinct forms in (1) the spatial organization of the urban economy, (2) the structures for social reproduction, and (3) the organization of the labor process. In these trends towards multiple forms of polarization lie conditions for the creation of employment-centered urban poverty and marginality and for new class formations.

The ascendance of the specialized services-led economy, particularly the new finance and services complex, engenders what may be regarded as a new economic regime, because although this sector may account for only a fraction of the economy of a city, it imposes itself on that larger economy. One of these pressures is towards polarization, as is the case with the possibility for superprofits in finance, which contributes to devalorize manufacturing and low-value-added services insofar as these sectors cannot generate the superprofits typical of much financial activity.

The superprofit making capacity of many of the leading industries is embedded in a complex combination of new trends: technologies that make possible the hypermobility of capital at a global scale and the deregulation of multiple markets that allows for implementing that hypermobility; financial inventions such as securitization which liquefy hitherto illiquid capital and allow it to circulate and hence make additional profits, the growing demand for services in all industries along with the increasing complexity and specialization of many of these inputs which has contributed to their valorization and often overvalorization, as illustrated in the unusually high salary increases beginning in the 1980s for top level professionals and CEOs. Globalization further adds to the complexity of these services, their strategic character, their glamour and therewith to their overvalorization.

The presence of a critical mass of firms with extremely high profit-making capabilities contributes to bid up the prices of commercial space, industrial services and other business needs, thereby making survival for firms with moderate profit-making capabilities increasingly precarious. And while the latter are essential to the operation of the urban economy and the daily needs of residents, their economic viability is threatened in a situation where finance and specialized services can earn superprofits. High prices and profit levels in the internationalized sector and its ancillary activities, such as top-of-the-line restaurants and hotels, make it increasingly difficult for other sectors to compete for space and investments. Many of these other sectors have experienced considerable downgrading and/or displacement, for example, the replacement of neighborhood shops tailored to local needs by upscale boutiques and restaurants catering to new high-income urban elites.

Inequality in the profit-making capabilities of different sectors of the economy has always existed. But what we see happening today takes place on another order of magnitude and is engendering massive distortions in the operations of various markets, from housing to labor. For example, the polarization of firms and households in the spatial organization of the economy contribute, in my reading, towards the informalization of a growing array of economic activities in advanced urban economies. When firms with low or modest profit-making capacities experience an ongoing if not increasing demand for their goods and services from households and other firms in a context where a significant sector of the economy makes superprofits, they often cannot compete even though there is an effective demand for what they produce. Operating informally is often one of the few ways in which such firms can survive, for example, using spaces not zoned for commercial or manufacturing uses, such as basements in residential areas, or space that is not up to code in terms of health, fire and other such standards. Similarly, new firms in low-profit industries entering a strong market for their goods and services may only be able to do so informally. Another option for firms with limited profit-making capabilities is to subcontract part of their work to informal operations.[14]

The recomposition of the sources of growth and profit-making entailed by these transformations also contribute to a reorganization of some components of social reproduction or consumption. While the middle strata still constitute the majority, the conditions that contributed to their expansion and politicoeconomic power in the postwar decades – the centrality of mass production and mass consumption in economic growth and profit realization – have been displaced by new sources of growth.

The rapid growth of industries with strong concentrations of high and low-income jobs has assumed distinct forms in the consumption structure, which, in turn, has a feedback effect on the organization of work and the types of jobs being created. The expansion of the high-income work force in conjunction with the emergence of new cultural forms has led to a process of high-income gentrification that rests, in the last analysis, on the availability of a vast supply of low-wage workers.

In good part, the consumption needs of the low income population in large cities are met by manufacturing and retail establishments, which are small, rely on family labor, and often fall below minimum safety and health standards. Cheap, locally produced sweatshop garments, for example, can compete with low-cost Asian imports. A growing range of products and services, from low-cost furniture made in basements to 'gypsy cabs' and family daycare, is available to meet the demand for the growing low-income population.

One way of conceptualizing informalization in advanced urban economies today is to posit it as the systemic equivalent of what we call deregulation at the top of the economy (see Sassen 1998: Chapter 8). Both the deregulation of a growing number of leading information industries and the informalization of a growing number of sectors with low profit-making capacities can be conceptualized as adjustments under conditions where new economic developments and old regulations enter in growing tension.[15] 'Regulatory fractures' is one concept I have used to capture this condition.

We can think of these development as constituting new geographies of centrality and marginality that cut across the old divide between poor and rich countries, and new geographies of marginality that have become increasingly evident not only in the less developed world but within highly developed countries. Inside major cities in both the developed and developing world we see a new geography of centers and margins that not only contributes to strengthen existing inequalities but also sets in motion a whole series of new dynamics of inequality.

The Localizations of the Global

Economic globalization, then, needs to be understood also in its multiple localizations, rather than only in terms of the broad, overarching macro-level processes that dominate the mainstream account. Further, we need to see that some of these localizations do not generally get coded as having anything to do with the global economy. The global city can be seen as one strategic instantiation of such multiple localizations.

Here I want to focus on localizations of the global marked by these two features. Many of these localizations are embedded in the demographic transition evident in such cities, where a majority of resident workers today are immigrants and women, often women of color. These cities are seeing an expansion of low-wage jobs that do not fit the master images about globalization yet are part of it. Their being embedded in the demographic transition evident in all these cities, and their consequent invisibility, contribute to the devalorization of these types of workers and work cultures and to the 'legitimacy' of that devalorization.

This can be read as a rupture of the traditional dynamic whereby membership in leading economic sectors contributes conditions towards the formation of strong labor aristocracy – a process long evident in western industrialized economies. 'Women and immigrants' come to replace the Fordist/family-wage category of 'women and children' (Sassen 1998: Chapter 5).[16] One of the localizations of the dynamics of globalization is the process of economic restructuring in global cities. The associated socioeconomic polarization has generated a large growth in the demand for low-wage workers and for jobs that offer few advancement possibilities. This, amidst an explosion in the wealth and power concentrated in these cities – that is to say, in conditions where there is also a visible expansion in high-income jobs and high-priced urban space.

'Women and immigrants' emerge as the labor supply that facilitates the imposition of low-wages and powerlessness under conditions of high demand for those workers and the location of those jobs in high-growth sectors. It breaks the historic nexus that would have led to empowering workers and legitimates this break culturally.

Another localization which is rarely associated with globalization, informalization, re-introduces the community and the household as an important economic space in global cities. I see informalization in this setting as the low-cost – and often feminized – equivalent of deregulation at the top of the system. As with deregulation (e.g., financial deregulation), informalization introduces flexibility, reduces the 'burdens' of regulation and lowers costs, in this case, especially the costs of labor. Informalization in major cities of highly developed countries – whether New York, London, Paris or Berlin – can be seen as a downgrading of a variety of activities for which there is an effective demand in these cities, but also a devaluing and enormous competition, given low entry costs and few alternative forms of employment. Going informal is one way of producing and distributing goods and services at a lower cost and with greater flexibility. This further devalues these types of activities. Immigrants and women are important actors in the new informal economies of these cities. They absorb the costs of informalizing these activities (see Sassen 1998: Chapter 8).

The reconfiguration of economic spaces associated with globalization in major cities has had differential impacts on women and men, on male and female work cultures, on male and female centered forms of power and empowerment. The restructuring of the labor market brings with it a shift of labor market functions to the household or community. Women and households emerge as sites that should be part of the theorization of the particular labor market dynamics evident today.

These transformations contain possibilities, even if limited, for the autonomy and empowerment of women. For instance, we might ask whether the growth of informalization in advanced urban economies reconfigures some types of economic relations between men and women? With informalization, the neighborhood and the household reemerge as sites for economic activity. This condition has its own dynamic possibilities for women. Economic downgrading through informalization creates 'opportunities' for low-income women entrepreneurs and workers, and therewith reconfigures some of the work and household hierarchies that women find themselves in. This becomes particularly clear in the case of immigrant women who come from countries with rather traditional male-centered cultures.

There is a large literature showing that immigrant women's regular-wage work and improved access to other public realms have an impact on their gender relations.

Women gain greater personal autonomy and independence while men lose ground. Women gain more control over budgeting and other domestic decisions and greater leverage in requesting help from men in domestic chores. Also, their access to public services and other public resources gives them a chance to become incorporated in the mainstream society – they are often the ones in the household who mediate this process. It is likely that some women benefit more than others from these circumstances; we need more research to establish the impact of class, education, and income on these gendered outcomes.

Besides the relatively greater empowerment of women in the household associated with waged employment, there is a second important outcome: their greater participation in the public sphere and their possible emergence as public actors. There are two arenas where immigrant women are active: institutions for public and private assistance, and the immigrant/ethnic community. The incorporation of women in the migration process strengthens the settlement likelihood and contributes to greater immigrant participation in their communities and *vis-à-vis* the state. For example, Hondagneu-Sotelo (1994) found immigrant women come to assume more active public and social roles, which further reinforces their status in the household and the settlement process. Women are more active in community building and community activism, and they are positioned differently from men regarding the broader economy and the state. They are the ones who are likely to handle the legal vulnerability of their families in the process of seeking public and social services for their families. This greater participation by women suggests the possibility that they may emerge as more forceful and visible actors and make their role in the labor market more visible as well. There is, to some extent, a joining of two different dynamics in the condition of women in global cities described above. On the one hand, they are constituted as an invisible and disempowered class of workers in the service of the strategic sectors constituting the global economy. This invisibility keeps them from emerging as whatever would be the contemporary equivalent of the 'labor aristocracy' of earlier forms of economic organization, when a low-wage worker's position in leading sectors had the effect of empowering that worker (i.e., the possibility of unionizing). On the other hand, the access to – albeit low – wages and salaries, the growing feminization of the job supply and the growing feminization of business opportunities brought about with informalization do alter the gender hierarchies in which they find themselves.[17]

The Global City: A Nexus for New Politicoeconomic Alignments

What makes the localization of the processes described above strategic, even though they involve powerless and often invisible workers, and potentially constitutive of a new kind of transnational politics is that these same cities are also the strategic sites for the valorization of the new forms of global corporate capital as described in the first section of this chapter.

Typically, the analysis of the globalization of the economy privileges the reconstitution of capital as an internationalized presence; it emphasizes the vanguard character of this reconstitution. At the same time it remains absolutely silent about another crucial element of this transnationalization, one that some, like myself, see as the counterpart of that capital, this is, the transnationalization of labor. We are still using the language of immigration to describe this process.[18] Secondly, that analysis overlooks the transnationalization in the formation of identities and loyalties among various population segments that explicitly reject the imagined community of the nation. With this come new solidarities and notions of membership. Major cities have emerged as a strategic site for both the transnationalization of labor and the formation of transnational identities. In this regard, they form a site for new types of political operations.

Cities are the terrain where people from many different countries are most likely to meet and a multiplicity of cultures can come together. The international character of major cities lies not only in their telecommunication infrastructure and international firms: it lies also in the many different cultural environments in which these workers exist. One can no longer think of centers for international business and finance simply in terms of the corporate towers and corporate culture at its center. Today's global cities are in part the spaces of post-colonialism and indeed contain conditions for the formation of a post-colonialist discourse (see, e.g., Hall 1991; King 1990).[19]

The large western city of today concentrates diversity. Its spaces are inscribed with the dominant corporate culture but also with a multiplicity of other cultures and identities. The slippage is evident: the dominant culture can encompass only part of the city.[20] And while corporate power inscribes these cultures and identities with 'otherness', thereby devaluing them, they are present everywhere. For example, through immigration a proliferation of originally highly localized cultures now have become presences in many large cities, cities whose elites think of themselves as 'cosmopolitan', or transcending any locality. An immense array of cultures from around the world, each rooted in a particular country or village, now are reterritorialized in a few single places, places such as New York, Los Angeles, Paris, London, and most recently Tokyo.[21]

Immigration and ethnicity are too often constituted as 'otherness'. Understanding them as processes whereby global elements are localized, international labor markets are constituted and cultures from all over the world are deterritorialized, puts them right there at the center of the stage, along with the internationalization of capital, as a fundamental aspect of globalization today. Further, this way of narrating the migration events of the postwar era captures the ongoing weight of colonialism and post-colonial forms of empire on major processes of globalization today and specifically those binding emigration and immigration countries.[22] While the specific genesis and contents of their responsibility will vary from case to case and period to period, none of the major immigration countries are innocent bystanders.

The centrality of place in a context of global processes engenders a transnational economic and political opening in the formation of new claims and hence in the constitution of entitlements, notably rights to place and, at the limit, in the constitution of 'citizenship'. The city has indeed emerged as a site for new claims: by global capital which uses the city as an 'organizational commodity' but also by disadvantaged sectors of the urban population, frequently as internationalized a presence in large cities as capital.

I see this as a type of political opening that contains unifying capacities across national boundaries and sharpening conflicts within such boundaries. Global capital and the new immigrant workforce are two major instances of transnationalized categories that have unifying properties internally and find themselves in contestation with each other in global cities. Global cities are the sites for the overvalorization of corporate capital and the devalorization of disadvantaged workers. The leading sectors of corporate capital are now global in both their organization and operations. And many of the disadvantaged workers in global cities are women, immigrants and people of color. Both groups find in the global city a strategic site for their economic and political operations.

The linkage of people to territory as constituted in global cities is far less likely to be intermediated by the national state or 'national culture'. We are seeing a loosening of identities from what have been traditional sources of identity, such as the nation or the village (Yaeger 1996). This unmooring in the process of identity formation engenders new notions of community of membership and of entitlement.

Yet another way of thinking about the political implications of this strategic transnational space is the notion of the formation of new claims on that space. Has economic globalization at least partly shaped the formation of claims?[23] There are indeed major new actors making claims on these cities, notably foreign firms who have been increasingly entitled to do business through progressive deregulation of national economies, and the large increase over the last decade in international businesspeople. These are among the new 'city users'. They have profoundly marked the urban landscape. Perhaps at the other extreme are those who use urban political violence to make their claims on the city, claims that lack the *de facto* legitimacy enjoyed by the new city users. These are claims made by actors struggling for recognition, entitlement, and for rights to the city.[24]

There is something to be captured here – a distinction between powerlessness and a condition of being an actor or political subject even though one lacks power. I use the term 'presence' to name this condition. In the context of a strategic space such as the global city, the types of disadvantaged people described here are not simply marginal; they acquire presence in a broader political process that escapes the boundaries of the formal polity. This presence signals the possibility of a politics. What this politics might be will depend on the specific projects and practices of various communities. Insofar as the sense of membership of these communities is not subsumed under the national, it may well signal the possibility of a transnational politics centered in concrete localities.

Conclusion

Large cities around the world are the terrain where a multiplicity of globalization processes assume concrete, localized forms. These localized forms are, in good part, what globalization is about. If we consider further that large cities also concentrate a growing share of disadvantaged populations – immigrants in Europe and the United States, African-Americans and Latinos in the United States, masses of shanty dwellers in the mega-cities of the developing world – then we can see that cities have become a strategic terrain for a whole series of conflicts and contradictions.

We can then think of cities also as one of the sites for the contradictions of the globalization of capital. On one hand, they concentrate a disproportionate share of corporate power and are one of the key sites for the overvalorization of the corporate economy; on the other hand, they concentrate a disproportionate share of the disadvantaged and are one of the key sites for their devalorization. This joint presence happens in a context where (1) the transnationalization of economies has grown sharply and cities have become increasingly strategic for global capital, and (2) marginalized people have found their voice and are making claims on the city as well. This joint presence is further brought into focus by the sharpening of the distance between the two.

These joint presences have made cities a contested terrain. The global city concentrates diversity. Its spaces are inscribed with the dominant corporate culture but also with a multiplicity of other cultures and identities, notably through immigration. The slippage is evident: the dominant culture can encompass only part of the city. And while corporate power inscribes non-corporate cultures and identities with 'otherness', thereby devaluing them, they are present everywhere. The immigrant communities and informal economy in cities such as New York and Los Angeles are only two instances.

The space constituted by the worldwide grid of global cities, a space with new economic and political potentialities, is perhaps one of the most strategic spaces for the formation of new, including transnational, types of identities and communities. This is a space that is both place-centered in that it is embedded in particular and strategic sites, and it is transterritorial because it connects sites that are not geographically proximate yet intensely connected to each other. It is not only the transmigration of capital that takes place in this global grid, but also that of people both rich (i.e., the new transnational professional workforce) and poor (i.e., most migrant workers) and it is a space for the transmigration of cultural forms, or the reterritorialization of 'local' subcultures. An important question is whether it is also a space for a new politics, one going beyond the politics of culture and identity, though at least partly likely to be embedded in these. The analysis presented in this chapter suggests that it is.

The centrality of place in a context of global processes engenders a transnational economic and political opening in the formation of new claims and hence in the constitution of entitlements, notably rights to place and ultimately, in the constitution of new forms of 'citizenship' and a diversity of citizenship practices. The global city has emerged as a site for new claims: by global capital which uses the city as an 'organizational commodity', but also by disadvantaged sectors of the urban population, frequently as internationalized a presence in large cities as capital. The denationalizing of urban space and the formation of new claims centered in transnational actors and involving contestation constitute the global city as a frontier zone for a new type of engagement.

Acknowledgement

This is an edited version of an article originally published in *American Studies*, 41:2/3 (Summer/Fall 2000): 79–95. We thank the journal.

Notes

1 Globalization is also a process that produces differentiation, but the alignment of differences is of a very different kind from that associated with such differentiating notions as national character, national culture and national society. For example, the corporate world today has a global geography, but it does not exist everywhere in the world: in fact, it has highly defined and structured spaces; secondly, it also is increasingly sharply differentiated from non-corporate segments in the economies of the particular locations (e.g., a city such as New York) or countries where it operates. There is homogenization along certain lines that cross national boundaries and sharp differentiation inside these boundaries.

2 We need to recognize the specific historical conditions for different conceptions of the 'international' or the 'global'. There is a tendency to see the internationalization of the economy as a process operating at the center, embedded in the power of the multinational corporations today and colonial enterprises in the past. One could note that the economies of many peripheral countries are thoroughly internationalized due to high levels of foreign investments in all economic sectors and of heavy dependence on world markets for 'hard' currency. What center countries have are strategic concentrations of firms and markets that operate globally, the capability for global control and coordination and power. This is a very different form of the international from that which we find in peripheral countries.

3 This proposition lies at the heart of my model of the global city (see Sassen 2001: Chapter 1).

4 More conceptually, we can ask whether an economic system with strong tendencies towards such concentration can have a space economy that lacks points of physical agglomeration. That is to say, does power, in this case economic power, have spatial correlates?

5 I see the producer services, and most especially finance and advanced corporate services, as industries producing the organizational commodities necessary for the implementation and management of global economic systems (Sassen 2000: Chapters 2–5). Producer services are intermediate outputs, that is, services bought by firms. They cover financial, legal, and general management matters, innovation, development, design, administration, personnel, production technology, maintenance, transport, communications, wholesale distribution, advertising, cleaning services for firms, security and storage. Central components of the producer services category are a range of industries with mixed business and consumer markets; they are insurance, banking, financial services, real estate, legal services, accounting and professional associations.

6 Telematics and globalization have emerged as fundamental forces reshaping the organization of economic space. This reshaping ranges from the spatial virtualization of a growing number of economic activities to the reconfiguration of the geography of the built environment for economic activity. Whether in electronic space or in the geography of the built environment, this reshaping involves organizational and structural changes.

7 Methodologically speaking, this is one way of addressing the question of the unit of analysis in studies of contemporary economic processes. 'National economy' is a problematic category when there are high levels of internationalization. And 'world economy' is a problematic category because of the impossibility of engaging in detailed empirical study at that scale. Highly internationalized cities such as New York or London offer the possibility of examining globalization processes in great detail within a bounded setting and with all their multiple, often contradictory aspects. King (1990) notes the need to differentiate the international and the global. In many ways the concept of the global city does that.

8   A methodological tool I find useful for this type of examination is what I call circuits for the distribution and installation of economic operations. These circuits allow me to follow economic activities into terrains that escape the increasingly narrow borders of mainstream representations of 'the advanced economy' and to negotiate the crossing of socioculturally discontinuous spaces.

9   This is illustrated by the following event. When the first acute stock market crisis happened in 1987 after years of enormous growth, there were numerous press reports about the sudden and massive unemployment crisis among high-income professionals on Wall Street. The other unemployment crisis on Wall Street, affecting secretaries and blue-collar workers was never noticed nor reported upon. And yet, the stock market crash created a very concentrated unemployment crisis, for example, in the Dominican immigrant community in Northern Manhattan where a lot of the Wall Street janitors live.

10   There is by now a vast literature documenting one or another of these various aspects (see generally Fainstein *et al.* 1993; see Abu-Lughod 1999 on New York, Chicago and Los Angeles, which she defines as the three US global cities).

11   Whether this has contributed to the formation of transnational urban systems is subject to debate. The growth of global markets for finance and specialized services, the need for transnational servicing networks due to sharp increases in international investment, the reduced role of the government in the regulation of international economic activity and the corresponding ascendance of other institutional arenas, notably global markets and corporate headquarters – all these point to the existence of transnational economic arrangements with locations in more than one country. These cities are not merely competing with each other for market share as is often asserted or assumed; there is a division of labor which incorporates cities of multiple countries, and in this regard we can speak of a global system (e.g., in finance) as opposed to simply an international system (see Sassen 2001: Chapters 1–4, 7). We can see here the incipient formation of a transnational urban system.

12   Further, the pronounced orientation to the world markets evident in such cities raises questions about the articulation with their nation-states, their regions, and the larger economic and social structure in such cities. Cities have typically been deeply embedded in the economies of their region, indeed often reflecting the characteristics of the latter – and they still do. But cities that are strategic sites in the global economy tend, in part, to disconnect from their region. This conflicts with a key proposition in traditional scholarship about urban systems, namely, that these systems promote the territorial integration of regional and national economies.

13   Elsewhere I have tried to argue that the current post-1945 period has distinct conditions for the formation and continuation of international flows of immigrants and refugees. I have sought to show that the specific forms of internationalization of capital we see over this period have contributed to mobilize people into migration streams and build bridges between countries of origin and the US. The implantation of western development strategies, from the replacement of small-holder agriculture with export-oriented commercial agriculture to the westernization of educational systems, has contributed to mobilize people into migration streams – regional, national, transnational. At the same time, the administrative commercial and development networks of the former European empires and the newer forms these networks assumed under the Pax Americana (i.e., international direct foreign investment, export processing zones, wars for democracy) have not only created bridges for the flow of capital, information and high-level personnel from the center to the periphery but, I argue, also for the flow of migrants (Sassen 1988). See also Hall's (1991) account of the postwar influx of people from the Commonwealth into Britain and his description of how England and Englishness were so present in his native Jamaica as to make people feel that London was the capital where they were all headed to sooner or later. This way of narrating the migration events of the

post-war era captures the ongoing weight of colonialism and post-colonial forms of empire on major processes of globalization today, and specifically those binding emigration and immigration countries. The major immigration countries are not innocent bystanders; the specific genesis and contents of their responsibility will vary from case to case and period to period.

14 More generally, we are seeing the formation of new types of labor market segmentation. Two characteristics stand out. One is the weakening role of the firm in structuring the employment relation; more is left to the market. A second form in this restructuring of the labor market is what could be described as the shift of labor market functions to the household or community.

15 Linking informalization and growth takes the analysis beyond the notion that the emergence of informal sectors in cities like New York and Los Angeles is caused by the presence of immigrants and their propensities to replicate survival strategies typical of Third World countries. Linking informalization and growth also takes the analysis beyond the notion that unemployment and recession generally may be the key factors promoting informalization in the current phase of highly industrialized economies. It may point to characteristics of advanced capitalism that are not typically noted. See Parnreiter *et al.* (1997) for an excellent collection of recent work focusing on the informal economy in many different countries.

16 This newer case brings out more brutally than did the Fordist contract, the economic significance of these types of actors, a significance veiled or softened in the case of the Fordist contract through the provision of the family wage.

17 Another important localization of the dynamics of globalization is that of the new professional women stratum. Elsewhere I have examined the impact of the growth of top-level professional women in high-income gentrification in these cities – both residential and commercial – as well as in the re-urbanization of middle-class family life (see Sassen 1998: Chapter 9; Sassen 2003).

18 This language is increasingly constructing immigration as a devalued process in so far as it describes the entry of people from generally poorer, disadvantaged countries, in search of the better lives that the receiving country can offer; it contains an implicit valorization of the receiving country and a devalorization of the sending country.

19 An interesting question concerns the nature of internationalization today in ex-colonial cities. King's (1990: 78) analysis about the distinctive historical and unequal conditions in which the notion of the 'international' was constructed is extremely important. King shows us how during the time of empire, some of the major old colonial centers were far more internationalized than the metropolitan centers. Internationalization as used today is assumed to be rooted in the experience of the center. This brings up a parallel contemporary blind spot well captured in Hall's observation that contemporary post-colonial and post-imperialist critiques have emerged in the former centers of empires, and they are silent about a range of conditions evident today in ex-colonial cities or countries. Yet another such blind spot concerns the possibility that the international migrations now directed largely to the center from former colonial territories, and neo-colonial territories in the case of the US, and most recently Japan (1994), might be the correlate of the internationalization of capital that began with colonialism.

20 There are many different forms such contestation and 'slippage' can assume. Global mass culture homogenizes and is capable of absorbing an immense variety of local cultural elements. But this process is never complete. The opposite is the case in my analysis of data on electronic components manufacturing, which shows that employment in lead sectors no longer inevitably constitutes membership in a labor aristocracy. Thus Third World women working in Export Processing Zones are not empowered: capitalism can work through difference. Yet another case is that of 'illegal' immigrants; here we see that national boundaries have the effect of creating and criminalizing difference. These

kinds of differentiations are central to the formation of a world economic system (Wallerstein 1990).

21   Tokyo now has several, mostly working-class concentrations of legal and illegal immigrants coming from China, Bangladesh, Pakistan or the Philippines. This is quite remarkable in view of Japan's legal and cultural closure to immigrants. Is this simply a function of poverty in those countries? By itself it is not enough of an explanation, since they have long had poverty. I posit that the internationalization of the Japanese economy, including specific forms of investment in those countries and Japan's growing cultural influence there have built bridges between those countries and Japan and have reduced the subjective distance with Japan (see Sassen 2000: 307–15).

22   The specific forms of the internationalization of capital we see over the last 20 years have contributed to mobilize people into migration streams (see Sassen 1988, 1998: Part One, and [13]). The renewal of mass immigration into the United States in the 1960s, after five decades of little or no immigration, took place in a context of expanded US economic and military activity in Asia and the Caribbean Basin. Today, the United States is at the heart of an international system of investment and production that has incorporated not only Mexico but areas in the Caribbean and Southeast Asia. In the 1960s and 1970s, the United States played a crucial role in the development of a world economic system. It passed legislation aimed at opening its own and other countries' economies to the flow of capital, goods, services and information. The central military, political and economic role the United States played in the emergence of a global economy contributed, I argue, both to the creation of conditions that mobilized people into migrations, whether local or international, and to the formation of links between the United States and other countries that subsequently were to serve as bridges for international migration. Measures commonly thought of as deterring emigration – foreign investment and the promotion of export-oriented growth in developing countries – seem to have had precisely the opposite effect. Among the leading senders of immigrants to the United States in the 1970s and 1980s have been several of the newly industrialized countries of South and Southeast Asia whose extremely high growth rates are generally recognized to be a result of foreign direct investment (FDI) in export manufacturing.

23   For a different combination of these elements see, for example, Dunn (1994).

24   Body-Gendrot (1999) shows how the city remains a terrain for contest, characterized by the emergence of new actors, often younger and younger. This is a terrain where it is the constraints and the institutional limitations of governments to address the demands for equity which engender social disorders. She argues that urban political violence should not be interpreted as a coherent ideology but rather as an element of temporary political tactics, which permits vulnerable actors to enter in interaction with the holders of power on terms that will be somewhat more favorable to the weak.

Chapter 5

A Question of Boundaries: Planning and Asian Urban Transitions

Michael Leaf

The very nature of urban planning practice requires that we should think carefully about the implications of boundaries. In the most fundamental sense of the term as it pertains to urban planning, boundaries refer to the basic parameters of problem definition, with the observation that what lies beyond may be insufficiently conceptualized, if not overlooked entirely. For practicing planners, the parameters of problem definition are shaped by a number of boundary conditions – jurisdictional responsibilities which lie at the root of planning as a spatial exercise, especially, but also the institutional limitations of prevailing regulatory regimes and the fiscal – and therefore temporal – boundaries of municipal budgeting.[1] Considering the possible disjuncture between the practice of planning and the processes and patterns of urbanization, especially rapid urbanization under conditions of limited societal resources, we need to ask whether the boundaries which define and delimit planning practice adequately fit with the overall parameters of urbanization, as understood spatially, temporally, and with regard to the institutional requirements of local governance.

As a step toward thinking about alternative means of addressing the rapid urbanization of developing countries in Asia, this chapter focuses on problems associated with the conventional boundaries we use in analyzing processes underlying urbanization and societal change and, by extension, in devising means for positive interventions, that is, planning and policy responses. Boundaries here refer to the spatial, temporal, social and regulatory parameters that are applied in thinking about and responding to the challenges of ongoing rapid urbanization. With reference to examples from China, Indonesia, and elsewhere among the rapidly urbanizing countries of the region, this chapter discusses 'the boundary question' with regard to three interlinked aspects of urbanization and urban change. At the broadest level of scale, how one understands the urban transition – described here as an outgrowth or manifestation of a society's demographic transition – is seen to be highly determined by the temporal and spatial parameters within which it is examined. Second, at the scale of the metropolitan region, the particular socio-spatial transformations that follow upon from the urban transition may be looked at with regard to a variety of boundary conditions, giving indications as to the increasingly problematic nature of newly urbanizing areas. And finally, understanding the conventional practice of planning as both a territorially bounded exercise and a predominantly statist undertaking – though with increasing linkages

89

to the growing capitalist sector – demonstrates how in many instances planning itself is part of the problem rather than a key to devising solutions. The challenge of planning in these terms also implies a need to rethink regulatory bounding as articulated in terms of informality or informalization, a persistent and perhaps growing phenomenon in the ongoing urban transitions of Asia.

Asia's Ongoing Urban Transitions

One particularly consequential example of the boundary problem is with regard to how we look upon two fundamental societal transitions that currently characterize much of the developing world: the continuing demographic transition and what may arguably be seen as its spatial correlate, the urban transition. The basic concept of the demographic transition derives from the understanding that the rate of natural change in any population will be determined by the combination of two other rates, (1) the rate at which new individuals are added (birth rate), and (2) the rate at which they are taken away (mortality rate). The demographic transition for any human society is best explained as the fundamental historical shift from an initial state of equilibrium (or near-equilibrium, such as a slow overall natural increase), where high birth rates are 'balanced' by high mortality rates, to one where another equilibrium plateau is reached where low mortality rates are again balanced by low birth rates. As it is a basic aspect of human nature to readily adopt those practices that prevent us from dying and to resist those which limit our ability to have children, the demographic transition in any society is played out over an extended period of time, triggered by an initial drop in mortality rates that is balanced out perhaps two or three generations later when birth rates exhibit a similar decline.[2] How one bounds one's temporal observation of such changes can be quite problematic, as perhaps most famously illustrated by the notion of the developing world's 'population bomb' as put forward in the 1960s, a polemical concept which can be interpreted as the result of an observer from a rich, post-transition society looking disparagingly at a slice through the middle of the ongoing transitions experienced by the poorer societies of the world.[3]

By comparison to the demographic transition, the urban transition may be defined as the shift within a society from a fundamentally agrarian existence, with perhaps only 10–15 per cent of its population resident in urban settings, to a point where that society has become predominantly urban, leveling off at an urban proportion of perhaps 75–85 per cent of the total population.[4] The links between these two transitions are complex and intricate, as we can see in this a blending of causes and outcomes. At one level, and with respect to more traditional notions of what constitutes 'urban', the urban transition may be interpreted as the spatial expression of the demographic transition, in that the demographic transition necessarily implies a fundamental shift in economic structure away from primary production (i.e., agriculture) to higher value-added forms of production historically associated with urban life. The two transitions are also linked by the basic adaptive psychology of the family and the position of children within the household economy as households shift out of agriculture and into urban spatial and economic settings. The usual argument is that children are fundamentally seen as assets in traditional rural agricultural settings, where their labor contribution can translate fairly directly into increased household productivity; in contrast, urban children are household liabilities, with long and costly periods of investment in education and preparation for joining the urban labor force. Societal shifts from rural to urban settings are thus seen over time to result in disincentives for large households, and urbanization, broadly speaking, is both a cause and an effect of the demographic transition.

In the various cases of Asia's ongoing urban transitions, the prospects appear daunting when one considers the total numbers involved. From Table 5.1, we can see the demographic indicators of Indonesia's urban transition from 1950 projected to 2030, expressed as both the urban proportion for the total national population, as well in the overall numbers of urban dwellers. Here we can observe the beginnings of a leveling off as Indonesia approaches its post-transition plateau, with urban dwellers reaching nearly two-thirds of the national population by 2030, up from less than 13 per cent in 1950, indicative of a significant drop in the per annum urban growth figures (to around 1.5 per cent, compared to a peak of more than 5 per cent in the 1980s). From the perspective of policy makers, these proportional figures are perhaps less critical than the overall totals of urban dwellers, who have increased markedly from less than 10 million in 1950 to more than 100 million today, a number likely to increase by another 75 million by 2030.

Table 5.1   Indonesian urban change, 1950–2030 (projected)

Year	Urban population (in 1000s)	Per cent urban (% p.a.)	Urban growth rate
1950	9863	12.40%	
1955	11633	13.46%	3.30%
1960	13996	14.59%	3.70%
1965	16840	15.79%	3.70%
1970	20501	17.07%	3.93%
1975	26047	19.36%	4.79%
1980	33377	22.20%	4.96%
1985	43552	26.15%	5.32%
1990	55819	30.59%	4.96%
1995	70357	35.60%	4.63%
2000	86943	40.99%	4.23%
2005	104048	46.17%	3.59%
2010	120986	50.90%	3.02%
2015	137554	55.01%	2.57%
2020	153006	58.42%	2.13%
2025	166792	61.12%	1.73%
2030	180069	63.66%	1.53%

Source: UN 2002.

Figures for China are comparable, as seen in Table 5.2, with a change in urban proportions from less than 13 per cent in 1950 to a projection of nearly 60 per cent by 2030. In the case of China, however, we can see the partial effects of national policy in the per annum urban growth rates of the 1960s and 1970s, as residential mobility controls constrained rural-urban migration and thus suppressed urban population growth. Economic reforms since the end of the 1970s have undermined the state's control on mobility, spurring a rapid increase in urban growth rates. Although the urban growth rates most likely peaked in the 1990s, and following demographic transition theory will level off over the next few decades (as with Indonesia, we may discern the beginnings of China's post-transition plateau by 2030), the daunting aspect again comes from the overall totals of urbanites in China, with an expansion from less than 70 million at the time of the founding of the People's Republic of China in 1949, to more than 500 million today. Again, most worrisome from the current perspective is the projected increase in urban populations by another 350 million over the next quarter century.

Table 5.2   Chinese urban change, 1950–2030 (projected)

Year	Urban population (in 1000s)	Per cent urban (% p.a.)	Urban growth rate
1950	69528	12.53%	
1955	86367	14.18%	4.34%
1960	105249	16.01%	3.95%
1965	128093	17.57%	3.93%
1970	144537	17.40%	2.42%
1975	161445	17.40%	2.21%
1980	196222	19.64%	3.90%
1985	246089	23.00%	4.53%
1990	316569	27.40%	5.04%
1995	382334	31.36%	3.78%
2000	456340	35.79%	3.54%
2005	535958	40.56%	3.22%
2010	617348	45.19%	2.83%
2015	698077	49.50%	2.46%
2020	771861	53.38%	2.01%
2025	834295	56.72%	1.56%
2030	883421	59.50%	1.14%

Source: UN 2002.

Regarding the policy implications of these overall trends, one can clearly see that there is a basic tendency to try to do those things that accelerate the demographic transition, that is, to develop interventions aimed at speeding up the drop in birth rates to offset the 'natural' effects of the drop in mortality rates that have accompanied modernization and economic change. The success in this regard of Indonesia's family planning program has made it exemplary for the developing world; and though certainly less voluntaristic, China's one child policy since the late 1970s has been no less consequential in demographic terms.

In contrast to such efforts to speed up demographic transitions, the overall orientation of policies addressing urban transitions may be argued to have the opposite effect, that is, to inhibit or at least control urban transitions by attempting to keep rural populations in place and delimiting or restricting the influx of rural migrants into metropolitan regions.[5] One can understand the predominant national level policy emphases on rural development as derived in part from cultural conservatism and a desire to address the needs of the countryside on what are considered to be its own terms, coupled with the understanding that reducing population pressures on urban areas serves the interests of existing urban populations and administrators. There are both ideological and practical considerations in the active suppression of urbanization through policy; not only are the challenges of rapid urbanization daunting for any administrative system, but notions of national identity for pre-transition societies tend to be rooted in ideals of rural agrarianism and are thus slow to change.

Accepting the argument that the demographic transition and the urban transition are two parallel and interlinked phenomena, or even two components of the same phenomenon, the observation that there are strikingly different policy reactions to these two transitions may appear contradictory. In fact, this is a very tangible expression of the boundary problem as I have described it. Policy-setting for shaping the demographic transition is undertaken for society overall – at the level of the nation-state – while policies to respond to the pressures of the urban transition are undertaken for the most part by much lower level jurisdictions, and hence are greatly shaped by the local politics of development. In policy terms, the urban transition is thus jurisdictionally bounded quite differently than the demographic transition.

A critical outcome of the localization of policy-making in response to rapid urbanization has been the persistence or reinforcement of what might be interpreted as variable or locally specific notions of citizenship, whereby entitlements to goods and services are determined in part by one's degree of administratively belonging to a specific locale – that is, being in the right place from the perspective of local jurisdictions. One pertinent example is seen in the decision in 1970 by Governor Ali Sadikin to declare Jakarta a 'closed city', limiting Jakarta residency to only those whose identity card was registered within the city.[6] On a more quotidian basis, one can identify other localized benefits of belonging in Indonesian cities, such as access to basic urban services and legal recourse in cases of land ownership conflict linked to specific registered categories of property ownership.[7] Parallel examples of locally specific citizenship in Chinese and Vietnamese cities are perhaps even more apparent, due to the persistent, though increasingly complicated, application of household registration systems put in place previous to the market reform periods in each country.[8]

Thinking about the policy responses to the region's urban transitions in contrast to how demographic transitions are addressed also raises concerns about the temporal bounding of policies regarding urbanization and development. There are distinct challenges to how one even conceives of the role of urban planning in such circumstances. What urban administrator or politician could conceive of proposing an eighty-year long term development plan for their city, let alone be able to develop this concept in a manner which could meaningfully address both the practical exigencies of ongoing rapid urbanization and the high level of uncertainty over such a long time frame?

Bounding Urban Spatial Transformations

Other sets of boundaries impinge upon our understanding of the socio-spatial changes associated with the urban transition at regional and local scales. I refer here first to jurisdictional boundaries that separate out older urban districts from more recently urbanizing rural and peri-urban settings, though this is only one aspect of boundary conditions within urbanizing regions. Such administrative boundaries may in some cases be reinforced and in other cases undermined by what I would term regulatory boundaries, which conceptually separate out formally planned or regulated areas from those where regulatory flexibility and unplanned development dominate. A third set of overlapping boundaries is what can be referred to as socio-spatial boundaries, whereby special development districts have been designated, either in older inner city settings for the purpose of planned redevelopment or on the urban edge for new development. By delimiting social access to sometimes large portions of the city, such boundaries serve to accelerate the growth of social segregation in Asian cities, giving spatial expression to the variable notions of citizenship that I referred to above.

Before looking in particular at this question of boundaries, it is useful to lay out some generalizations regarding processes and patterns of urbanization in the developing countries of Asia – clearly a risky undertaking considering the variety of particular circumstances across the cities of the region. However, one may generalize the urbanization process in the region as resulting from the intersection of two broad sets of socio-spatial change, the inward movement of erstwhile rural populations as agricultural labor is shed from the countryside, and the outward expansion of urban spatial economies and urban populations following upon new urban class formation and the growing influence of market forces in the allocation of urban space. It is not just incidental that both sets of changes are occurring in concert with expanding globalization, as the financial, technological, institutional and ideological changes associated with transborder flows are consequential in shaping the spatial outcomes of urban change.

To briefly summarize the spatial transformations arising from the basic necessity of creating cities that can accommodate expanded proportions of national populations through the course of the region's urban transitions, we may distinguish between changes in pre-existing urban form through inner city redevelopment and the formation of new urban spaces at the outward edge of expanding metropolitan regions.

Inner City Redevelopment

Major inner city changes arise for the most part through planned transformations which are often couched in political symbolic terms with regard to the need for 'modernizing' the city, creating the 'City of the 21st Century' and so forth, or in other words, linking to the developmentalist rhetoric of the state, whether or not changes are directly undertaken by the state itself. The broad set of outcomes we can observe in this regard include the following:

- Increasing spatial separation by land use and the dissolution of traditional, or at least long-standing, patterns of mixed use;
- Formation of specialized inner city districts, typically argued for not only in terms of modernization but in the interest of accommodating the needs of international capital;
- Loss of traditional urban built forms and character, a point of deep frustration for those interested in heritage conservation, or even for those who promote the development of new built forms referential to local cultural and environmental factors;
- Residential deconcentration out of inner cities, coupled with the perhaps surprising observation that as cities grow, their inner city populations are often seen to be decreasing (and arising from this, a growing distinction between day-time and night-time populations, due to the increasing separation of work and housing, production and consumption);
- Rapid growth of transportation needs and problems related to separation of land uses, as manifested in increasing traffic congestion, underinvestment in public transportation (in many cases, even development of basic road networks), and the related decline of urban air quality, among other environmental impacts.

Such inner city changes are dependent upon and an integral part of the local rise of capital through the formation of often large-scale real estate development companies, the 'big business' of urban development. These structural changes in urban economies are manifested differently in the various institutional contexts of the region, as, for example, in the formation of land development companies associated with banks and trading conglomerates in Thailand, the Philippines and Indonesia, or through local government-related companies created by municipal, district, and sub-district agencies in Chinese and Vietnamese cities – seen as a form of 'red capitalism' in China in the recent past, but which has become less and less red as time goes on. There are important transborder implications here as well, which might be labeled as 'regionalization' if not 'globalization', as is apparent from the sources of investment for major inner city redevelopment projects in Ho Chi Minh City and Hanoi, which are dominated by Korean, Taiwanese and Singaporean capital sources.[9]

Peri-urban Development

Socio-spatial changes arising from the outward expansion of metropolitan regions is an area of investigation which has so far received relatively limited research attention (and planning responses) compared to inner city changes – a major shortcoming, certainly, if one considers how important such changes are and will continue to become as the region's urban transitions proceed apace. As an exception, one can point to the work of McGee *et al.* (1995) who have worked on the formation of 'Extended Metropolitan Regions' (EMR) in Asia, arguing for the need to examine change as a form of 'region-based' urbanization, in contrast to more traditional tendencies to understand urbanization as a 'city-based' phenomenon.[10] In thinking about the growing phenomenon of the EMR in Asia, one can point to three basic factors that are shaping such changes:

- Traditionally high rural population densities, this is a condition often associated with wet-rice agriculture, which creates the possibility for 'urbanization in place' across a wide swath of territory as urban economies extend outward into the countryside. Population densities can influence industrial location decisions, as factories can be situated far afield to take advantage of low labor costs of the erstwhile countryside. Workers are able to subsist on low wages, as many of their costs of household reproduction are still covered through household agricultural production, and many processes can be cheaply outsourced through proximate cottage industries. In this we see parallels to the new logic of 'post-Fordist' production, although the formation of 'post-peasant' households[11] do not necessarily fit with the creation of 'learning regions' as envisioned by theorists.
- Technological changes which facilitate the rapid outward expansion of urban economies. Here one might first think of the rapid growth of telecommunications into the countryside. Perhaps more critical, however, is what has been referred to in Indonesia as the 'colt revolution', in reference to the Mitsubishi Colt, typical of the small minivans which have become ubiquitous for the rapid and inexpensive movement of people and goods throughout urban regions, connecting inner city locales with settings deep into the countryside. The rapid improvement of local roads and transportation options over the course of the reform era has also been seen to greatly enhance the connectivity between rural China and the country's urban system.[12]
- Weakness of regulatory controls across wide territories. Erstwhile rural localities are often overwhelmed with the speed, scale and newness of these changes. The pre-existing lack of local regulatory capacity is often complicated as well by the increased interjurisdictional competition between peri-urban localities for investment and development. In such contexts, one can find very localized instances of the 'race to the bottom' as local administrations allow substandard construction, weak enforcement of environmental regulations and concessionary labor practices in their efforts to secure investments that will facilitate the transition out of agriculture for local residents. In the case of China, such outward expansionary effects were compounded in the early reform period by policies to promote what were referred to as Township and Village Enterprises (TVEs) in the interest of keeping peasants out of cities by creating viable local livelihood alternatives (under the slogan of 'leaving the land but not the village').[13]

Looking at such fundamental and far-reaching urban transformations across the region, one finds a variety of ways by which jurisdictional, regulatory and socio-spatial boundaries overlap and intersect. Within Indonesian cities, for example, the often fine-grained differentiation of land parcels by legal status underpins regulatory differences within urban districts, with weaker property right claims characterizing what are disparagingly seen as *kampung* neighborhoods cheek by jowl with more highly regarded – and more legally secure – registered parcels.[14] In looking at the effect of changes within a metropolitan region, studies of the peri-urbanization of Jakarta have shown how manipulation of the regulatory system by formal sector developers has in many cases sidelined the administrative functions of local peri-urban jurisdictions and disenfranchised rural landowners in the formation of what is essentially an extensive private sector land bank surrounding the city.[15] In the case of Chinese cities, the long-standing legal and administrative distinctions between urban and rural lands, with urban lands considered state lands under municipal authority while village committees maintain authority over designated rural lands, has created a wide regulatory grey area for peri-urban development. Administrators in the villages surrounding Chinese cities have been able to enter into agreements for a range of types of land development in the conversion of their agricultural lands, from small scale and often sub-standard residential and industrial development, to larger scale, though still 'informal' developments,[16] to the creation of properly planned residential compounds and industrial estates on village territories.[17] The important point of commonality across all such cases is the ascendancy of market forces, regardless of the degree to which they engage the interests of local administrators and whether or not they lead to speculative or directly productive outcomes.[18]

Regulatory Bounding and the Role of Planning

The discussion so far has dealt with what might be termed 'interpretive' or 'analytical' uses of boundaries as they pertain to thinking about the ongoing urban transition in Asia, that is, the delineation of categories, territories, and time frames that are conventionally used in seeking to understand the processes through which urbanization occurs. In this concluding section, I will focus more specifically on questions of regulatory boundaries and how they shape the practice of urban planning in the region.

Urban planning practice, by its nature, is a spatially bounded activity. The implicit assumption of so much of planning is that development or urban socio-spatial change is 'targetable', that is, that a particular spatially distinct group – a community, a village, a neighbourhood, a city, or a region – is to be the recipient of the benefits of a planned intervention. Hence, one speaks of neighborhood planning, community planning, regional planning, and so forth. Such tendencies toward spatial bounding are problematic in any context, though perhaps even more so in contexts of rapid urban change, as planning in this conventional sense is not able to fully capture the spatial fluidity of the urban transition or to respond to the growing trans-localism of migrant households and the rapid changes in social make-up of growing peri-urban settlements. One might therefore consider the growing disjuncture between the effects of 'natural development', that is, social and economic change or betterment without conscious planned intervention, and what is envisioned through official or formal processes of planned urbanism.

Perhaps even more critical than the spatial bounding of planning practice for shaping the future of urban societies in the midst of rapid urbanization are the regulatory and institutional settings within which planning practice is situated. In practice, the regulatory bounding of planned interventions is so often undermined by the flexibility of regulatory practices on the ground, with much of what previously has been termed 'Third World urbanization' characterized by 'informality' or the informalization of local regulatory regimes.[19] 'Informality' in this sense applies to a broad swath of administrative actions, from the exigencies of local administrators responding to the particularistic needs of their constituents, even when such actions may be contradictory to policies established at higher administrative levels,[20] to the flexible application or manipulation of development requirements arising from collusion and conflicts of interest, if not outright corruption, between private developers and local officials.[21]

The persistence or continuing reproduction of informality in the rapidly growing cities of the erstwhile developing world may at one level be explained in terms of local capacities, with the recognition that existing human and institutional resources are simply insufficient for dealing with the speed and scale of urban change. And though one would not want to disparage ongoing efforts within the international community for promoting 'capacity building' in the fields of urban planning and management, such efforts need to be contextualized in terms of broader structural factors in order to fully engage the challenges of regulatory informality.[22] Beyond the political economy implications of looking at informality in this manner, one may also identify the complex effects of evolving governance practices in fostering the reproduction of informality, practices which may be understood in terms of potentially highly localized cultures of governance.[23]

Tensions between local authority structures deriving from persistent local cultures of governance and the formally structured rules of the game as inherent in institutionalized planning practice are apparent throughout the growing cities of developing Asian countries. Perhaps the most extreme response to the challenge of modernization through planned urbanization in contexts of limited resources and rapid socio-spatial change is seen in the many instances of what I would term the 'new town impulse' which seems to be at the heart of urban modernism throughout the region.

The new town impulse has become a signal characteristic of Chinese urbanism in the reform era, driven by the deconcentration of urban populations into newly formed residential enclaves and fostering the rapid outward spread of cities throughout much of the country's highly populated coastal regions. This impulse to urban modernism is also apparent in large-scale real estate developments from Muang Thong Thani on the northern outskirts of Bangkok, to Putrajaya adjacent to Kuala Lumpur, to Saigon South on the southern border to Ho Chi Minh City, to the stalled project for Citra West Lake City to the northwest of Hanoi.[24] Perhaps the most extreme – or egregious – impact of the new town impulse has been in the metropolitan region surrounding Jakarta, where the highly concentrated and politically well-connected real estate development industry has managed to administratively capture large tracts of territory through their manipulation of the Indonesian development permit system, essentially creating a massive private sector land bank with sufficient capacity to accommodate the productive output of the industry for decades to come.[25]

In contexts where persistent regulatory informality constrains the wholesale transformation of pre-existing cities into modern urban centers built to 'international standards' or otherwise adhering to globalized notions of urban modernism, the new town impulse may be understood as the attempt to limit the spatial scope of planning and thus achieve what is possible on a reduced scale. If the vision of a properly planned and regulated urban environment is unattainable for the burgeoning Asian metropolis overall, the rationale behind the new town impulse argues for carefully delimiting the boundaries of the planned city in order to create environments conducive to the attraction of foreign investment and the accommodation of growing segments of middle-class – or at least consumer-class – urban residents in formally regulated settings. In this way, the formal planning and construction of new residential districts accelerates the socio-spatial fragmentation of the Asian city by accentuating the social exclusivity of broad swaths of urban territory. Newly established administrative boundaries accompanying new town development thus reinforce both the boundaries of regulatory formality and the socio-spatial boundaries of class segregation, in contrast to the mixed patterns of socio-spatial integration in older urban forms. If one accepts the notion that socioeconomic segregation is a deleterious phenomenon in that it exacerbates tensions between groups and limits opportunities for social advancement, the role of planning in these instances should be recognized as being socially regressive insomuch as it facilitates such trends.

As I have argued above, these patterns of urban transformation and the limited though consequential role that formal planning processes have played are an outgrowth or response to broad social trends associated with the demographic transitions ongoing throughout the region, transitions which are linked to one extent or another to the multiple transborder flows subsumed under the aegis of globalization – flows of technologies, production processes and institutional forms, among others. Globalization figures into these transformative changes in the urban landscape in other ways, perhaps most crucially in the globally expanding logic of capitalism as the basis for socioeconomic (and socio-spatial) development. The growth of domestic capitalism, spurred as it has been by the rapidly expanding values of urban spatial economies across the region, has linked into, and thus reconfigured, long-standing traditions of statist development. Despite the urban landscapes of fragmentation and social segregation that have resulted, the accumulative desires of the development industry in concert with the ideological impulses of state developmentalism will no doubt continue to shape urban spatial transformations across the region for some time to come.

Notes

1   A similar observation can be made concerning the temporal, spatial and institutional limitations to the conventional 'project' approach to community level planning interventions. For a critique of how such forms of practice fail to engage the 'politics of patience' necessary for working with urban poor communities, see Appadurai (2001).

2   The factors that have been examined as initial triggers to historic demographic transitions cover a diverse range, including among others: scientific and technical advances; the advent of the germ theory of disease; introduction of new cultigens and consequent

dietary and health improvements; institutional change, including the development of the nation-state and the consequent expansion of entitlements, such as the growth of universal childhood education; overall economic change, most significantly the shift out of subsistence agriculture and the expansion of capitalist production; and from an ecologist's point of view, the significant shift in energy flows, derived from the fossil fuel revolution – in short, a series of factors which might be broadly subsumed under the rubric of modernization. For an overview of the basic concepts underlying demographic transition theory and its current application, see Chenais (1992) and Jones et al. (1997), respectively; see also Ness (2000) for analysis of Asian urbanization trends from this perspective.

3   The principle reference here is to Ehrlich (1968). For a particularly acerbic critique of Ehrlich's reading of the demographic transition, see Bookchin (1994).

4   As examples of post-transition societies whose urban proportions are more or less leveling off, one could look at the US, with 77.2 per cent of its national population living in cities as of 2000, the UK, with 89.5 per cent, Canada, with 78.7 per cent, or Japan with 78.8 per cent (UN 2002).

5   Although strategies of rural development in general, as well as specific approaches derived from growth pole theory, may be seen to have had only limited success in diverting population growth and investment away from major urban centres, such approaches persist nonetheless in the practice of regional development planning. For critical reviews of the field, see Higgins and Savoie (1997) and Douglass (1998).

6   Although enforcement of this decree has never been successful, it has yet to be repealed. See Abeyasekere (1989) for more on the policies of the Ali Sadikin administration and their effects on the development of Jakarta.

7   For overviews of the complexity of the land right situation in Jakarta and its implications for tenure security and overall patterns of urban development, see Leaf (1993; 1994).

8   Local variation in the entitlements of citizenship is a central theme in Solinger's (1999) work on 'contesting citizenship' by rural migrants in Chinese cities. In such contexts of administrative differentiation of classes of citizens (between rural and urban citizens in China through the *hukou*, or household registration system, or in the parallel *hokhau* system in Vietnam), the argument can be put forward that such formalized administrative categories reinforce rather than create distinctions between urban and rural populations, and between residents of differing regions of the country.

9   See Leaf (1999) for a discussion of the effects of such regionalization of investment in Hanoi.

10   For regional overviews and variations, see the papers assembled in Ginsburg, Koppel and McGee (1991) and McGee and Robinson (1995). For more specific treatments in China, see Guldin (2001) and Marton (2000) and in the Philippines, see Kelly (2000).

11   See Rigg (2003) for a discussion of the term and its applicability across the changing rural landscape of Southeast Asia.

12   Guldin (2001) in particular provides examples of this localized form of 'space-time compression' in the case of southern China, where a trip from village to city has been reduced from three days, to one day, to ninety minutes, over the course of less than two decades.

13   Marton (2000) emphasizes how the most successful such enterprises have tended to be located within emerging extended metropolitan regions, and Webster (2001) discusses how the shake-out over time, whereby the most competitive TVEs have been able to expand while the majority have collapsed, has resulted in the phenomenon of peri-urban rustbelts on the edge of many of the most vibrant cities in China.

14   The fine-grained nature of socio-spatial patterns has long characterized Indonesian urbanism, with elite and lower classes living in close proximity. For discussion of the social implications of such patterns and how they are changing over time, see Guinness

(1986) on Jogjakarta and Abeyasekere (1989) on Jakarta. The complexity of land ownership claims in Jakarta is detailed in Leaf (1993).

15 For details on the land regulatory system in Jakarta and how it has been manipulated, see Leaf (1991) and Cowherd (2002).

16 The most famous example of such a large scale informal settlement is that of 'Zhejiang Village' on the outskirts of Beijing, built by small-scale developers from southern Zhejiang Province in collaboration with local Beijing villagers, which at its height housed over 100,000 rural migrants and extensive garment production facilities for domestic and international markets. See Zhang (2001) for a detailed history and analysis of this settlement.

17 Further examples are given in Leaf (2002).

18 Considering the implicit high profits to be derived from land and property development under conditions of rapid urbanization in concert with the growth of market economies, it is not surprising to find high levels of speculative activity in peri-urban zones. Examples may be seen in the work of Cowherd (2002) on the extreme overbuilding of luxury housing in Jakarta, and in the case of China in the analyses of 'zone fever' by Cartier (2001) and other manipulations of peri-urban land markets by Ho and Lin (2003).

19 This stream of analysis originated out of the debates around the 'informal sector' since the 1970s (Moser 1984) though with a specific emphasis on the role of state regulation in determining informality (Portes, Castells and Benton 1989). See Leaf (2005a) for a recent overview of the topic.

20 See Koh (2000) for a detailed analysis of this aspect of urban local governance in a Vietnamese setting.

21 Analyses of such systematic informality in the Jakarta property development sector are found in Leaf (1991) and Cowherd (2002).

22 And here one could point to the arguments of Sassen (2000) and other world systems theorists (see Tabak and Crichlow 2000) that the continuing growth in informality, that is, processes subsumed under the heading of informalization, is indicative of a cyclical phase in the ongoing expansion of the world economy.

23 This is an argument developed in Leaf (2005b) in the context of Chinese urban governance; for parallels in Indonesia, see Cowherd (2002) on the local administrative implications of the 'culture' of modernization and development.

24 An analysis of the planning and urban development implications of the Citra West Lake project is given in Leaf (1999).

25 Estimates of the scale of this land bank relative to the annual output of the formal real estate development industry indicate that land sufficient for somewhere between 70 and 150 years of housing construction had been administratively captured by developers by 1995 (Cowherd 2002: 279; see also Winarso 1999 and Firman 2004).

Chapter 6

Shifting Drivers of Change, Time-space Telescoping and Urban Environmental Transitions in the Asia-Pacific Region

Peter J. Marcotullio

Cities in the Asia-Pacific region have undergone rapid and extensive transformations over the past few decades. These changes have been linked to regional and global economic flows (Lo and Yeung 1996). Flows of trade, investments, people and information have impacted urban environmental conditions as development trajectories have been substantially altered (see, e.g., Burgess *et al.* 1997; Lo and Marcotullio 2001). Globalization flows have been accompanied by rapid economic and population growth within cities of the region. Over the last 50 years, the Asia-Pacific region has more than doubled the percentage of people living in cities from 17 to 38 per cent, meaning the addition of 1.15 billion people. Together these influences along with the rapid diffusion of new technologies have had dramatic impacts on the urban environmental conditions within the region.

Interestingly, however, a popular theory in policy circles describes the relationship between development and the environment as an 'inverted-U shaped' path where environmental impacts and economic growth first have a positive correlation, and after reaching a specific level of development, the relationship reverses. This theory, known as the Environmental Kuznets Curve (EKC), suggests that all countries and cities are on one and the same path. It is supported by common descriptions of the environmental challenges of Third World cities as similar to those already experienced by the developed world. For example, a 1998 *The Economist* article on development and the environment starts out with a description of the environmental problems in English cities of the mid-1800s written by Friedrich Engels (*The Economist* 1998) suggesting that things are very similar now in developing cities to conditions in Manchester, UK, in the 19th century. Further, the notion is supported by literature suggesting that while cities in different parts of the world might have been different in the past, they are becoming more alike. Michael Cohen (1996) suggests that cities of the North and South are converging in physical form and problems. Within studies of the Asia-Pacific region many other scholars agree (see, e.g., Dick and Rimmer 1998).

This chapter argues that contemporary forces, including globalization flows, have impacted Asia-Pacific urban environmental transitions in a specific and unique manner. The accumulation of changes brought upon by drivers has brought with it unprecedented challenges defined by shifts in timing, speed and sequencing of previously experienced development patterns such that they occur sooner, faster and more simultaneously. The process is called 'time-space telescoping'. The chapter elaborates on the driving forces associated with the emergence of the process and its impacts.

To fulfill these tasks, the chapter is divided into four sections. The first section presents the theory used to conceptualize the changing nature of the development and environment relationship. This section includes an introduction to urban environmental transition theory and the elaboration of time and space effects that result in 'time-space telescoping'. The second section examines the shifts in some of the drivers of environmental change. Attention is directed to economic growth, demographic changes and shifts in technologies. The third section analyses the changes within nations and cities by presenting evidence for time-space telescoping. Examination of environmental conditions and processes within the region demonstrate that the transitions are occurring sooner, faster and more simultaneously than previously experienced.

Conceptualizing Time-space Effects

The Urban Environmental Transition Theory and the Notion of Shifting Environmental Challenges

In a recent text that sums up 10 years of research in this field, Gordon McGranahan *et al.* (2001) present a persuasive argument concerning the relationship between development, affluence and the urban environment. Their claims are that urban environmental burdens tend to be more dispersed and delayed in more affluent settings than in poorer cities. Dividing the impact of environmental problems into categories, they suggest that as cities move from poverty to wealth, their environmental burdens shift from localized, immediate and health threatening to global, delayed and ecosystem threatening.

The model is powerful for three reasons. First, it allows for a distinction amongst environmental burdens. Sets of environmental challenges, including those in the 'brown', 'gray', and 'green' agendas are grouped together in terms of their emergence within development paths and their impacts. Second, wealth is used as the central axis to define shifts between these sets of burdens. While each set of burdens has a different relationship to increasing wealth, the ordering remains clear within the model. Lastly, injecting the issue of geographic and temporal scale in the model helps to incorporate very different types of problems previously dealt with separately. Within the urban environmental transition theory, lack of water supply has a relationship to global greenhouse gas emissions. Urban activities impact larger areas with growing wealth. As a city moves beyond the 'brown' agenda, for example, environmental impacts of cities increase in scale from the household and neighborhood levels to citywide regions. For those cities struggling with the 'green' agenda, the dominant environmental impacts of urban-based activities are regional if not global.

The model suggests the brown agenda issues arise first. They are predominately found in poor and growing cities and include environmental health and local issues relating to inadequate water and sanitation, urban air quality and solid waste disposal. 'Gray' agenda issues are associated with industrialization and urbanization, including chemical pollution of the air and watersheds. The 'green' agenda issues are those associated with ecological sustainability and resource degradation. Impacts related to these areas are dominant among wealthy cities, which are the largest contributors to global environmental burdens and other largely extra-urban problems including greenhouse gas emissions, ozone depleting substances, acid rain, persistent organic pollutants, among other concerns.

The urban environmental transition lays the ground for exploring how these various sets of burdens emerge within the rapidly developing world context and how transitions between brown, gray and green agendas have been transformed by contemporary drivers.

Time-space Telescoping

The contemporary development context, as compared to the past, has been impacted by both time- and space-related effects. Time-related effects include changes in the speed in which human activities occur, including movement, work, learning, etc. Time-related effects speed up and intensify our lives' daily experiences. They are often associated with technological improvements.

Space-related effects include processes that concentrate activities in geographically uneven patterns. Space-related effects create diversity in social and physical landscapes. The concentrations may be of economic activity, of people – in cities – or specific types of people (e.g., poor people in slums).

Importantly, these effects have changed over human history. Geographers and urban planners have identified a number of influences shaping urban environments that are related to time and space effects and historical urban geographers have identified a number of factors that have had differing impacts over time (Carter 1983). Recent work on globalization suggests time- and space-related effects through the concentration of certain types of infrastructure (e.g., communication, transportation, financial and business services, headquarters of transnational companies) in specific locations around the world (e.g., world cities) (Friedmann 1986; Friedmann and Wolff 1982; Lo and Yeung 1998; Sassen 1991) and through new transportation and communications technologies which have transformed the ability of information, goods and services to move across space (see, e.g., Dicken 1998).

Scholars have been studying these two effects as processes shaping the constraints placed upon human activities by both time and space and thus, how socioeconomic drivers impact human activities. Janelle (1968, 1969), for example, identified the increasing speed at which people move across space and its affects on economic activity and social relations. His term, 'time-space convergence', defines the process of decreasing the friction of distance between places and results in decreases in the average amounts of time needed to travel between them. Processes creating 'time-space convergence' have not only made the world smaller, but also have increased our ability to impact a larger number of different environments around the world, at a more intense rate. Harvey (1989) focused on what he called 'time-space compression' as underpinning the emergence of the

post-modern condition. Time-space compression includes processes that
revolutionize the objective qualities of space and time and alter, sometimes in quite
radical ways, how we represent the world to ourselves. Smith (1990) and Smith and
Lee (1993) have identified changing patterns of risk over time, as traditional risks
– those associated, for example, with local indoor air pollution – have now
combined with more modern risks (e.g., those associated with inhaling pesticides
sprayed on biofuels) and generated a new genre of mixed risks in low-income
countries. These notions demonstrate changes in the speed and location of
activities over time, which have thus transformed human behaviors and their
resultant environmental impacts.

Another time-space effect that shapes the contemporary relationship between
development and the environment in the developing world is time-space
telescoping. This results in the collapsing, compression and telescoping of
previously experienced sequential development patterns so that they occur sooner,
faster and more simultaneously (Marcotullio 2004). The difference this process has
made on environmental transitions is demonstrated in Figure 6.1. As cities are now
developing under strong pressures of globalization and are further influenced by a

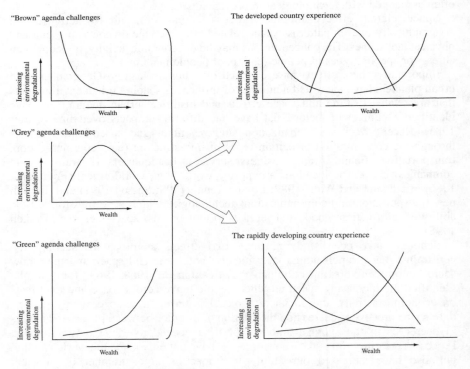

**Figure 6.1   Comparison of environmental transition experiences between
developed and rapidly developing countries**

Table 6.1   National 1990 real GDP for selected Asian countries and comparative real GDP and year for the USA

Country	1990 per capita GDP	Year of similar US GDP	US GDP during year
Japan	18,789	1979	18,789
Hong Kong	17,491	1977	17,567
Singapore	14,258	1967	14,330
Taiwan	9,910	1950	9,561
South Korea	8,704		
Malaysia	5,131	1913	5,310
Thailand	4,645		
Indonesia	2,516	1870	2,445
Philippines	2,199		
China	1,858		
Vietnam	1,040	1820	1,257
Cambodia	945		
Laos	933		
Burma	751		

Source: Maddison, A. (2001) *The World Economy: A Millennial Perspective*, Paris: OECD, Table A1-c, p. 185; Table A3-c, p. 215; Table C1-c, p.279; and Table C3-c, pp.304–306.

variety of demographic and technological shifts, there are many goods and services available to them that were not previously available to the developed world at similar levels of income. Thus, environmental burdens associated with resource consumption, waste and pollution are found increasingly at lower levels of income. Further, these impacts are rising faster, as the technologies are more available, and their consumption is aided by rapidly increasing numbers of middle and upper-income households. At the same time, however, there remain a substantial portion of the population within these cities with much less income and accessible services. These two sets of conditions are found side-by-side, demonstrating an overlapping or telescoping of burdens.

In order to appreciate how these conditions are different from what was previously experienced, comparisons of conditions within the contemporary rapidly developing world must be made to conditions experienced by the current industrialized world when the latter had incomes similar to those now reached by the former. Table 6.1 lists the 1990 level of GDP per capita in a variety of countries in East and Southeast Asia and compares them with the dates when they were reached by a developed country, the United States. If environmental conditions remained similar at levels of income throughout time, we would expect to be able to compare the conditions in these countries to those of the USA in the 19th to early 20th century. Table 6.1 forms the basis of the rationale for the analyses presented later in the chapter.

Shifting Drivers of Change: Wealth, Population and Technology

Driving forces are those actions that change nature from its conditions independent of humankind (Kates *et al.* 1990). The term has been increasingly used in the integrative sciences (see, e.g., Sustainable Science Initiative, Resilience Alliance network, etc.)[1] where it has been further characterized in a number of ways (primary versus proximate, anthropogenic versus bio-physical, dependent versus independent, primary versus secondary).

While more sophisticated analyses have been applied to the identification and analysis of drivers, the simplified model used in this chapter, introduced by Ehrlich and Holdren (1971), remains both useful and powerful. That model, called IPAT, equates human pressure or impacts on the environment (I) to the demand on the earth's resources (A – affluence), times the number of people (P – population), times the impact per unit of resource (T – technology). Explicitly, the model equates increasing environmental impact with increases in both affluence and population, but technology may either raise or lower environmental burdens – depending upon the efficiency of the technology used.[2] What this section attempts is the examination of changes in these drivers comparing those in the late 19th century and early 20th century with those now. That is, at the turn of the century, the now industrialized world was at similar levels of economic development that the now developing world has reached. The question that this section addresses is: Given approximately similar levels of development, what were the differences in driving forces?

Affluence: Economic Growth

In the context of the IPAT model, affluence is translated as per capita consumption. Per capita consumption is known to rise with GDP. It has been the experience of most nations and individuals that consumption increases with economic growth. Moreover, as historical comparable statistics for economic growth rates are available, their comparison helps to define the context within which environmental transitions occur (Maddison 2001).

The current rates of national economic expansion are different from those of the past. The speed of economic development is faster (Crafts 2000). Among developing countries, those in the Asia-Pacific region have experienced the most rapid changes and these development rates are unprecedented. Table 6.2 compares annual rates of growth by decade between selected Asia-Pacific region countries and both established industrialized countries and newly industrialized ones in Latin America. These data demonstrates that during decades of rapid growth for the United States and some other European nations, the rates of increase did not match those of the now rapidly developing world. In the past, growth rates of more than 4 per cent were rare. The rates of rapid growth experienced by countries in the Asia-Pacific region, among others, are historically new.

Underpinning these growth rates is a number of qualitatively and quantitatively new processes. Many of these processes are related to economic globalization flows. Some of the important flows include those involving trade, investment, people, knowledge and information. While these flows occurred in the past, the contemporary era is defined as being driven by globalization processes (Dicken

Table 6.2   Annual growth of GDP per capita in selected countries, by decade (percentages)

Country	1870–80	1880–90	1890–1900	1900–10	1910–20	1920–30	1930–40	1940–50	1950–60	1960–70	1970–80	1980–90
Australia	1.90	0.40	-1.04	2.64	-1.00	-0.52	2.17	1.97	1.69	3.14	1.72	1.75
Austria	1.11	1.63	1.66	1.33	-3.06	4.05	0.99	-0.66	5.81	4.11	3.53	1.92
France	1.23	1.15	1.93	0.30	0.85	3.46	-1.14	2.69	3.65	4.46	2.63	1.73
Germany	0.83	2.02	2.13	1.19	-1.65	3.09	3.19	-2.55	7.05	3.50	2.56	1.97
Italy	0.53	0.54	0.68	2.71	1.05	1.21	1.85	-0.01	5.39	5.09	3.25	1.99
UK	0.86	1.43	1.14	0.26	0.55	0.42	2.34	0.45	2.27	2.24	1.80	2.47
USA	2.65	0.62	1.89	1.95	1.13	0.42	1.21	3.15	1.58	2.87	2.09	1.81
Canada	0.61	2.73	2.04	3.40	-0.51	2.22	1.10	3.31	1.84	3.35	3.31	1.87
Japan	0.99	1.76	1.54	1.00	2.66	0.88	4.50	-3.82	7.55	9.31	3.33	3.53
South Korea	na	na	na	1.10	2.10	-0.03	3.54	-6.13	4.04	5.42	6.39	8.14
Taiwan	na	na	na	0.45	1.49	1.86	1.78	-3.52	4.26	6.76	7.66	6.24
Indonesia	na	na	1.17	2.10	0.59	2.18	-0.60	-2.59	2.61	0.92	4.20	3.05
Philippines	na	na	–	3.22	na	na	-0.44	-1.45	1.41	1.73	3.55	-0.85
Thailand	na	na	0.29	0.41	na	na	0.41	0.19	1.95	4.49	4.09	5.76
China	na	na	0.59	0.54	na	na	-0.01	-2.34	3.64	2.21	2.96	6.33
Argentina	na	na	2.50	3.32	-0.95	1.62	0.20	1.83	1.09	2.76	1.22	-2.23
Brazil	na	na	-0.92	1.22	1.66	1.25	2.07	2.54	3.39	2.76	5.51	-0.86
Mexico	na	na	1.57	2.18	-0.75	0.30	1.27	2.97	2.92	3.10	3.36	-0.50

Source: Calculations from data providing from Maddison (1995): Monitoring The World Economy 1820–1992, Table D–Ia, pp. 194–206

Notes: Japan data for 1880 is 1885

China, Indonesia, Philippines, South Korea, Taiwan and Thailand data for 1910 are for 1913

China, Indonesia, Philippines, South Korea, Taiwan and Thailand data for 1930 are for 1929

China, Indonesia, Philippines, South Korea, Taiwan and Thailand data for 1940 are for 1938

'–' data are not available

na: data not available

Managing Urban Futures

1998, Held *et al.* 1999, Johnston *et al.* 1995, Knox and Agnew 1998). The result has been the deepening, thickening and speeding up of economic, social and political interdependencies. Despite the 1997–98 Asia-Pacific region financial crisis, the region has been increasingly tied into the global economic system and has continued to expand economically.

An example of the difference between the experiences of the rapidly developing world and now developed world can be partially seen in Table 6.3. Growth in trade over the past few decades among countries in the region has been spectacular. Scholars agree that world growth in both total volume of trade and rates of increase are larger than in previous periods. Moreover, growth among the rapid

Table 6.3   Annual average trade export growth rates, by decade (percentages)

Country/Region	1950–60	1960–70	1970–80	1980–90	1990–2000
World	6.4	9.2	20.4	6.1	6.6
USA	5.5	8.1	18.5	5.7	7.3
UK	4.7	5.9	18.5	5.8	5.4
France	6.0	9.7	20.3	7.5	4.2
Germany	16.2	11.2	19.1	9.2	3.9
Netherlands	9.4	11.0	19.8	6.5	5.5
Italy	10.5	13.9	20.0	8.7	4.6
Australia	0.9	7.7	15.9	6.3	4.0
Japan	15.9	17.5	20.8	8.9	4.1
Korea	1.3	39.8	37.2	15.1	10.1
Hong Kong	−0.3	14.5	22.4	16.8	8.3
Taiwan	6.5	23.2	28.6	14.8	7.2
Singapore	−0.1	3.3	28.2	9.9	9.9
Malaysia	0.6	4.2	24.2	8.6	12.2
Indonesia	−1.1	1.6	35.3	−0.3	8.1
Thailand	1.7	5.9	24.7	14.0	10.5
Philippines	5.0	5.4	18.4	3.9	18.8
China	18.1	1.3	20.0	12.9	14.5
Vietnam	−4.7	−22.7	64.9	18.9	22.7
Cambodia	2.4	−1.6	−10.1	11.4	26.2
Lao PDR	11.3	29.6	−17.7	10.3	15.4
Myanmar	1.8	−8.8	14.7	−7.9	14.4
Argentina	−0.3	4.8	18.0	2.1	10.1
Brazil	−2.1	7.2	21.8	5.1	5.9
Venezuela	8.1	1.2	20.3	−4.6	5.5
Mexico	3.1	6.1	24.8	8.2	16.1

Source: UNCTD (1999): *Handbook of International Trade and Development Statistics* New York, United Nations, Table 1.5 and 1.6, pp. 14–21. UNCTD (2002): *Handbook of Statistics* New York, United Nations, Table 1.2A, pp. 16–23.

Table 6.4   FDI inflows, by host region and economy, 1990 to 2001, in billions of US$

Host region/Economy (per cent of total)	1990–95 (Annual) Average	1996	1997	1998	1999	2000	2001	Cumulative growth 1997–2001	Average FDI inflow 1997–2001
World	225.3	386.1	478.0	694.4	1,088.3	1,491.9	735.1	4,487.7	897.5
Developed economies	145.0	219.9	267.9	484.2	837.8	1,227.5	503.1	3,320.5	664.1
European Union	84.2	110.4	127.9	262.2	487.9	808.5	322.9	2,009.4	401.9
Other Western European Countries	3.2	5.5	9.9	12.5	19.3	23.5	13.2	78.4	15.7
North America	47.0	94.0	114.9	197.2	307.8	367.5	151.9	1,139.3	227.9
Other Developed economies	10.6	9.9	15.1	12.2	22.7	27.9	15.0	92.9	18.6
Developing countries	74.3	152.7	191.0	187.6	225.1	237.9	204.8	1,046.4	209.3
Africa	4.3	5.8	10.7	9.0	12.8	8.7	17.2	58.4	11.7
Latin America and the Caribbean	22.2	52.8	74.3	82.2	109.3	95.4	85.4	446.6	89.3
Asia	47.3	93.3	105.8	96.1	102.8	133.7	102.0	540.4	108.1
Central Asia	662.0	2.6	3.8	3.1	2.5	1.9	3.6	14.9	3.0
South, East and South-East Asia	44.6	87.8	96.3	86.2	99.9	131.1	94.4	507.9	101.6
The Pacific	388.0	663.0	150.0	277.0	229.0	88.0	198.0	942.0	188.4
Central and Eastern Europe	6.0	13.5	19.1	22.6	25.4	26.5	27.2	120.8	24.2

Source: UNCTAD, FDI/TNC database. *World Investment Report 2002*, New York: United Nations, Table B.1, pp. 303–306.

industrializers of the Asia-Pacific region is consistently larger than that of total average world trade. Foreign direct investments (FDI) have grown at even faster rates than trade, reaching almost US$1.5 trillion in 2000 before falling in 2001 and 2002 (see Table 6.4). In 2000, developing East and Southeast Asia alone captured 8.8 per cent of total global FDI and 55 per cent of the FDI into developing countries (United Nations Conference on Trade and Development 2002). These are only a few examples of how both economic growth and the processes underlying them are significantly different in the Asia-Pacific region than experienced by already industrialized countries at similar levels of income.

Demographic Shifts

Demographic change has been described as the 'first tier driving force of environmental change throughout the history of mankind' (Kates *et al.* 1990). Despite arguments over whether population growth is the cause of environmental damage (Ehrlich and Ehrlich 1990) or the 'ultimate resource' (Simon 1981), scholars agree that it is a driver of change.

The size and structures of populations in the developing world are different from those of the developed world at similar levels of income. A brief history of human population growth helps to point out these differences. Estimates suggest that around 1700, there were approximately 700 million people living on the planet (Meyer 1996). Between 1750 and 1800 North America, Europe and Russia led the world's population expansion. Between 1850 and 1900 Latin America joined these leaders of population growth. Both Asia and Africa fell below world average rates (China was ravaged by war, disaster and famine and the population did not grow at all during the last half of the 18th century). By 1900, there were approximately 1.65 billion people on the planet; 56 per cent lived in Asia, 18 per cent in Europe, 8 per cent in the former Soviet Union, 5.5 per cent in North America, 4.5 per cent in Latin America, and less than 0.5 per cent in Oceania (Demeny 1990). During the years of rapid western economic growth (1900–1950) populations increased from 300 to 400 million in Europe (0.7 per cent growth rate)[3] and from 90 to 170 million in the USA (1.2 per cent annual average growth rates). While growth in the North America was extremely rapid in its early periods, growth rates were largely due to starting from very low absolute levels. In the words of Demeny (1990), the region was a 'virtual demographic vacuum in 1700'.[4]

Within contemporary Asia-Pacific region countries, the number of people living and moving to cities has dwarfed those of previous times. If one includes the 18 economies defined by the UN as part of East and Southeast Asia as the Asia-Pacific region,[5] the region reached a total urban population of 147 million, which made up of 17 per cent of the total regional population in 1950. By 2000, the number had grown to over 760 million, doubling its population share to 38 per cent. By 2030, the number of people living in Asia-Pacific region cities is expected to double. The UN predicts that 1.3 billion people will be urban accounting for approximately 55 per cent of the total population in the region (UN 1999).[6]

A comparison of urbanization rates demonstrates the differences between the development paths of these countries. Figure 6.2 suggests that for some countries in the Asia-Pacific region, urbanization occurred at much more rapid rates than it did for the United States. While the UK urbanized at a slower rate than that of the US,

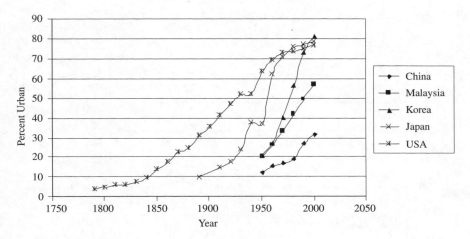

Figure 6.2   Comparative urbanization

the fastest urbanizers in the world are the Japanese and Koreans. Countries such as Malaysia and China are not far behind. China is expected to achieve urbanization levels of close to 50 per cent by 2030.

As a note of caution, urbanization growth rates should not be confused with individual city growth rates. Historically, individual cities in the developed world have grown as fast as or even faster than contemporary cities of the Asia-Pacific region. For example, New York City at the turn of the 19th century experienced extremely rapid growth rates, as it was a port of call to the United States during a period of heavy and rapid immigration. The difference now is that many or most of the cities in the developing world are experiencing this type of growth – and not only from immigration – as opposed to only a few. The current era of development is one defined by the emergence of large cities (Lo and Yeung 1998) and most of these large cities are located in the developing world and most within Asia.

Population structures have also changed much more quickly in the currently rapid developing Asia-Pacific region than those in the now industrialized countries. One example of this change is the share of the 'aged' population within countries. Typically, an aged society is one when 7 per cent or more of the population is 65 years or older (i.e., the elderly population). The growth of this share of the population has been related to postponed fertility, because in economically more mature societies, parents put off having children and have less of them. In the UK and Germany, the 7 per cent share was achieved in 1930 after these countries reached a level of approximately G-K$5500 and G-K$4000, respectively.[7] In the United States, the aged population did not reach these levels until the beginning of the 1940s when it reached G-K$7000, and Japan reached the mark in 1970 (Miura 2000) at the GDP per capita level of G-K$9700.

Within rapidly developing Asia-Pacific region countries, however, there is a rapidly growing elderly population. In China, for example, an estimated 6.8 per cent

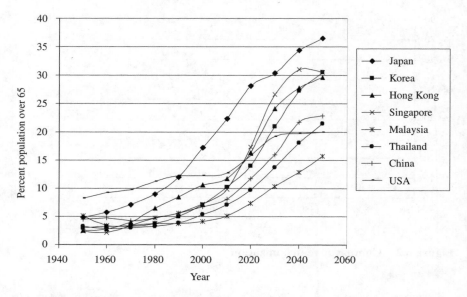

Figure 6.3   Growth of urban population over 65

of the population was 65 years or older by 1999 (Yoshida and Ma 1999),[8] at which point it reached a comparative economic level of G-K\$3200. Moreover, many of the rapid industrializers in the region are predicted to increase their aged population shares much more rapidly than, for example, the United States (see Figure 6.3).

Another important trend is the decrease in household sizes. While this trend is appearing around the world (Liu *et al.* 2003) those in the developing world are growing in size at much lower levels of income than those in the developed world. In China, a country traditionally known for large families, household size dropped to 3.97 in 1990 (Yoshida and Ma 1999) and continues downward. The spatial aspects of these trends are important to such issues as biodiversity, as many of the 'hotspots' are located near areas of increasing numbers of households and the associated increasing demands on local resources have the potential to impact them substantially.

Technological Change

Technology has been defined as 'a means whereby humans use nature for their own benefit' (Headrick 1990). This chapter views shifts in technologies through the lens of long waves of economic development theory proposed by Nikolai D. Kondratieff (Kondratieff 1979). Using available data on Germany, France, England and the USA, Kondratieff demonstrated, among a group of variables,[9] a secular trend in a specific direction that was structurally linked to the overall changes in the economic environment of the particular society. While Kondratieff studied three types of waves of varying length (long ones of approximately 50 years duration, middle ones of seven to 10 years duration and short ones of three to four years), he was primarily interested in the longest waves, which ultimately took on his name. These waves were characterized by accelerating rates of price increases from deflationary depression to inflationary peaks, followed by decade-long plunges from the peaks to primary troughs, which again are followed by recovery. There is debate over the underlying reasons why these exist (see, e.g., Maddison 1991)[10] but the general consensus is that long-term fluctuations in national economies of some 50 years duration have occurred. That is, industrialization was not a process of gentle increases of production over time, but was cyclical, driven by spurts of activity followed by periods of slower or no growth.

Associated with these spurts of activity are shifts in technologies. Brian Berry (1997) suggests that USA history, for example, is marked by a rise and fall of a succession of techno-economic systems, defined by interrelated sets of technologies with associated sets of raw materials, sources of energy and infrastructure networks. The first set of techno-economic systems involved the use of wind, water and wood. This powered the cotton mills of England at the beginning of the 19th century. In the middle of the century, the use of coal, steam and iron spread throughout Europe. Steam was first used to produce energy on demand and used first in pumping then in turning machines and then in transportation vehicles such as boats, ships and railroads. In the third phase, steel, kerosene and electricity appeared. Improved processes changed steel from a costly and rare product to a cheap and useful building material. In the fourth phase, the technologies underpinning growth included petroleum, internal combustion and chemicals. Some believe we are now in a fifth phase driven by the production of electronic devices (e.g., computers and television) and the emergence of service-related technologies (education, entertainment, defense, finance, etc.) (Headrick 1990).

The addition of technologies to the long wave theory is important in that it provides an underlying rationale for such changes. Indeed, as the argument is pursued, each system opens up novel products and factor markets that in turn produce major surges of economic growth. Onto these different eras of development can be mapped shifts in urban environmental challenges and for most of those in the United States, their solutions. Solutions, in this sense, often meant the export, both geographically and temporally (i.e., the future), of the particular set of environmental problems. At the source of these problems, however, for that time and those places, conditions improved.[11] Table 6.5 attempts to summarize the environmental challenges faced by cities in the United States during the different eras associated with techno-economic long wave cycles.

The way that technological advances have been secured has also changed. At first the creation of technological innovations was generated by trial and error, produced by craftsmen and taught by the traditional apprenticeship system. By the end of the century, however, curiosity and technical difficulties turned crafts technology into engineering and then into science. Nowadays, science is intermingled with technology at every level, and education has become a prerequisite for almost every technical craft (Headrick 1990). These differences are important to contemporary development patterns.

Table 6.5 Techno-economic systems, urban development eras and associated urban environmental challenges

Kondratieff Cycle	Technology systems	Urban development era	Environmental challenges
To 1845 A-Phase (peak 1814) B-Phase (trough 1843)	Wind, water, wood	Mercantile cities	Health (small pox, yellow fever, convulsions, cholera) water supply, street refuse, domestic animals, street drainage
From 1845 to 1895 A-Phase (peak 1864) B-Phase (trough 1896)	Coal, steam and iron	Early industrial cities	Health (pneumonia, tuberculosis, diarrhea) high densities, horses and rapid urban growth, provision of adequate water
From 1895 to 1945 A-Phase (peak 1920) B-Phase (trough 1930s)	Steel, kerosene and electricity	National industrial cities	Health (pneumonia, tuberculosis, diarrhea, occupational), clean water, mass consumption, sanitation, sewerage, solid waste beginnings of air pollution
From 1945 – Present A-Phase (peak 1970s) B-Phase (?)	Petroleum, internal combustion engines and chemicals	Mature/post-industrial cities	Health (circulatory system, cancer, nephritis, pneumonia including influenza) non-point source of air and water pollution cumulative persistant chemical and toxic waste (including nuclear waste) consumption related issues (land use, energy consumption, GHGs emissions) emerging pollutants (PM2.5)

Notes:
Kondratieff Cycles: Kondratieff 1979; Yeates 1998, (citing Mager 1987) p. 64.
Technological systems: Berry 1997, p. 302; Lo 1994.
Urban phases: Yeates 1998; Melosi 2000.
Environmental challenges: health related from Jackson 1995; water supply, sewerage and sanitation from Melosi 2000; Chudacoff 1981; other from Melosi 2000; Hays 1987, various sources.

The means of transportation and communications play a role in every aspect of the transformation of the earth and a key role in both time and space effects. Transportation technologies are keys to the expansion of both industry and agriculture during the 19th and 20th centuries as well as to the separation of consumption from production and to globalization in general. Moreover, our relationship to space has been transformed through the promotion of the automobile and airplane. The efficiency and effectiveness of the transmission of goods, services and knowledge across geographical space has improved, making these items available at lower costs, in greater quantities and across a larger geographic span than ever before (Drucker 1986).[12] Cities within the Asia-Pacific region have encouraged improvements in their transportation systems (see, e.g., Lo 2000). In some cases, they are ahead of the world. Shanghai, for example, has the only MAGLEV train in operation.

Recent technologies related to information and communications (ICT)[13] have changed the character of both business and social activities. At the same time, they have been unevenly distributed, thus having inequitable effects on cities (Alaedini and Marcotullio 2002). Many of the cities in the Asia-Pacific region have been quick to pick up on these technologies, often supported by supply side mega-infrastructure projects and have therefore benefited in terms of growth and international linkages. For example, Thailand has more cellular phones than all of Africa and the installation of fiber optics in the 'intelligent corridor' along the major arterial ring road of Bangkok by the government reinforces the linear expansion of the city into exurban areas (Hack 2000). The Malaysian Multimedia Super Corridor (MSC), estimated at US$20 billion, is a giant urban planning initiative and the heart of the national development strategy. Singapore's efforts to equip the export-oriented manufacturing enclaves of Johor and the Riau Islands, conceived of as being independent industrial townships, with infrastructure connections will allow these facilities to directly communicate with the city with state-of-the-art telecommunications (United Nations Centre for Human Settlements 2001).

Other Drivers

The above presented examples are not the only drivers responsible for change. Bennett and Dahlberg (1990) point to the importance of institutions related to the rules of property, exchange and regulation of the environment. These influences shape the way that affluence, demography and technology are applied in different societies. Moreover, the Millennium Ecosystem Assessment (2003), a global association of scholars assessing the ecosystem services trends and conditions of the planet, suggests a number of other direct and indirect drivers. In their framework they identify five general categories of indirect drivers including demographic, economic (including globalization, trade and market policy frameworks) and sociopolitical factors, science and technology, and cultural and religious values. These are the drivers that work at diffuse levels, difficult to measure and observe. More direct drivers are included in the categories changes in local land use and land cover, species introduction and removal, technology adaptations and use, external inputs (e.g., fertilizers, pest controls, etc.), harvest and resource consumption, climate change and natural physical and biophysical drivers (e.g., volcanoes, evolution, etc.) uninfluenced by humans. While these are all important to specific situations, including those discussed in this chapter, they can only be mentioned and not directly examined. No doubt, some play a role in facilitating time-space telescoping.

Impacts of Time-space Telescoping on Asian Cities

The impacts of these shifting drivers are many and diverse, but a specific set is addressed in the theoretical section of the chapter. Hypothetically, given the differences in driving forces described, a comparison of environmental conditions between countries and cities of the developed and rapidly developing world, using the notion of 'time space telescoping', would result in three related findings:

- Contemporary rapidly developing countries and cities have experienced some urban environmental pressures at lower levels of income than those experienced by industrialized countries;
- At similar income ranges, rapidly developing countries and cities have experienced faster growth in environmental pressures than those experienced by industrialized countries;
- Rapidly developing countries and cities have experienced sets of environmental pressures more simultaneously than developed countries and cities.

Adequately testing these claims requires an enormous amount of standardized data across time and for individual cities, which is not currently available. At the same time, some anecdotal evidence is immediately on hand. There is evidence, for example, that environmentally related processes are emerging at lower levels of income than previously experienced. Urbanization, in general, is occurring sooner. Countries in the Asia-Pacific region are urbanizing at lower levels of income than the United States (see Figure 6.4). For countries such as Japan, Korea, Malaysia and Indonesia, urbanization began at lower levels of income, and at any given GDP level

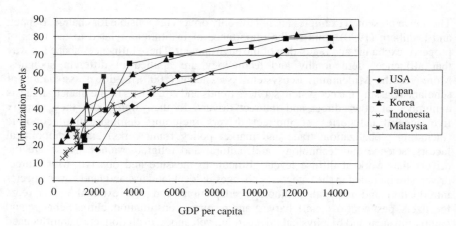

Figure 6.4   Comparison of urbanization levels and GDP per capita

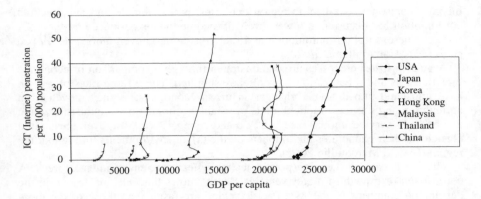

Figure 6.5   Comparison of growth in ICT (Internet) penetration

similar to that of the USA, they have a higher urbanization level. Also, new technologies are being adopted much sooner. For example, in rapidly developing countries ICT diffusion, such as Internet use and cell phone adoption, are increasingly found at high penetration levels at lower levels of income than experienced by developed countries (see Figure 6.5). When comparing motor vehicle ownership levels between cities in the Asia-Pacific region, it is clear that those of lower income cities (such as Bangkok and Kuala Lumpur) are similar if not greater than those of established industrialized cities (such as Tokyo) (see Table 6.6). This has translated into higher fuel consumption at lower levels of

Table 6.6   Vehicle ownership and per capita income

City	Car ownership (per 1000 persons)	Motorcycle ownership (per 1000 persons)	Vehicle ownership (per 1000 persons)	1990 Per capita income (US$)
Tokyo	236	36	272	39,953
Hong Kong	46	4	50	14,101
Singapore	110	42	152	12,939
Seoul City	123	18	201	5,942
Kuala Lumpur	206	201	407	4,066
Bangkok	220	220	399	3,826
Jakarta	92	113	205	1,508
Manila	79	18	87	1,099

Source: Barter 1999, Kenworthy et al, 1997.

income. For some Asia-Pacific region countries, motor vehicle and aviation fuel
consumption has occurred at lower levels of income than experienced by the USA
(an extreme example of automobile and fuel consumption even among developed
countries) (Marcotullio *et al.* 2004).

Arguably, faster processes are demonstrated through a comparison of economic
growth levels and urbanization rates already supplied. These rapid rates of growth
have translated into dramatic changes in urban economies, as they moved from
agricultural and services-oriented mode of wealth accumulation to industrialization
and business services modes (see, e.g., Lo and Yeung 1996). Moreover, these rapid
changes in their economies had impacts on the urban environments throughout the
region (Lo and Marcotullio 2001).

The same can be seen in specific technological and environmental areas. A
comparison of growth in motor vehicles clearly shows that rates of growth in the
developing countries of the Asia-Pacific region are faster than those of the more
developed countries. For example, Barter (1999) demonstrated that from 1960 to
1993, some Southeast Asia cities (Bangkok, Kuala Lumpur, among others) had
experienced faster growth rates of motor vehicle registration per 1000 inhabitants –
and ultimately reached higher levels per capita – than Tokyo, Singapore and Hong
Kong, despite having much lower income levels.

Urban landscapes are also changing faster than previously experienced. Urban
land use changes have been dramatic in rapidly developing Asia-Pacific cities. One

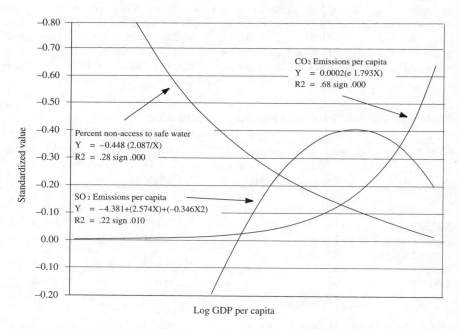

**Figure 6.6   Percentage of non-access to safe water, CO_2 and SO_2 emissions
per capita**

example is the movement of the title for tallest buildings in the world from the developed to the developing world. Taipei, Taiwan now has the world's tallest building, Taipei 101, completed in 2004, but other tall buildings can be seen in Kuala Lumpur, Hong Kong and Seoul. Massive and rapid development of tall buildings in Shanghai's Pudong area is also unprecedented.

In terms of evidence for the more simultaneous emergence of sets of environmental challenges, we compare the history of the now industrialized world with the current situation in rapidly developing cities. An extensive literature on the history of urban development, and on urban technology and infrastructure in the United States demonstrates the sequential manner within which these changes took place (see, e.g., Chudacoff 1981; Melosi 1999; Melosi 2000; Tarr 1978, 1996, 1999). When viewing the current situation in the now developing world, things appear differently. Figure 6.6 maps three proxies for the different agenda challenges. The 'brown' agenda is represented by the curve for non-access to water supplies. The 'gray' agenda of the industrialization and motorization era is represented by the curve for SO_2 and the 'green' agenda is represented by the curve for greenhouse gas emissions. The figure demonstrates the overlapping nature of these issues in a cross section of national economies. When the percentages of the total world urban population are estimated from the intersections of these curves, the results demonstrate that more than 50 per cent of those living in cities are experiencing simultaneous multiple burdens (see Table 6.7).

Table 6.7   Estimated urban population living under various environmental conditions, 1995

1995 GDP Category (US$)	Environmental challenge	Total urban population (thousands) (N)	Share of total
<467.74	Lack of water and sanitation ('brown' issues)	456,985	17.8%
>467.75 and <1,071.52	Rising industrial pollution ('gray' issues), and significant 'brown' issues	518,812	20.3%
>1,071.53 and <3,981.07	High 'grey' issues, rising modern risks ('green' issues) and 'brown' issues	526,315	20.6%
>3,981.08 and <14,125.3	High but decreasing 'gray' issues, rising 'green' issues	296,993	11.6%
>14,125.3	Largely 'green' issues	613,480	24.0%
Missing		147,610	5.8%
Total global urban population		2,560,195	

Source: Marcotullio P.J.; Rothenberg S. & Nakahari M. (2003). Globalization and urban environmental transitions: Comparison of New York's and Tokyo's experience. *Annals of Regional Science* 37: 369–390.

To test the simultaneity of emerging environmental concerns demands a comparison of the histories between cities or between specific environment conditions. One study suggests that while New York City has undergone its environmental transitions in a more sequential manner, Tokyo's transitions were compressed (Marcotullio *et al.* 2003). Specifically, starting in the 1970s, Tokyo was battling both those challenges related to the brown agenda, demonstrated by the need to build its sanitation infrastructure, as well as challenges related to industrialization and motorization, demonstrated in reductions in air and water pollution levels. Another study compares the growth of transportation systems between the USA and Asia-Pacific countries, noting that while the transitions between stages of systems within the USA are clearly visible, they are much less so in East and Southeast Asian cities (Marcotullio and Lee 2003). While there is much to be done to further test these claims, evidence is mounting in their support.

Conclusions

This chapter presented the theory of time-space telescoping along with the driving forces that created it and some examples of its impacts. The focus was on countries that are rapidly developing and specifically on those in the Asia-Pacific region. Globalization processes and other driving forces have been experienced differentially.

Besides geographic limitations, there are also other necessary qualifications to the findings. In the space provided, only the skeletal frame of work in this area could be presented. Much empirical work is needed to confirm, refine or refute these ideas.

The theory's value is that it identifies some of the important aspects of the new context for development and the environment, particularly within cities and always for an understanding of current trends. To summarize the theory suggests that drivers of change – those examined in this chapter include economic development, technological advances and demographic shifts – are significantly different in the rapidly developing world than when the now developed world was industrializing. This simple fact is often missed. Indeed, despite the massive changes that have been documented elsewhere, theories based on the Environmental Kuznets Curve (EKC) persist.

The change in drivers has had specific impacts on environmental transitions, although according to the theory, these transformations have impacts on other spheres of urban development as well. Initial comparisons confirm the expectations that for rapidly developing countries and cities environmental challenges crop up sooner in the development process, increase faster over time, and emerge in a more simultaneous order than previously experienced by the developed world. Some data has been presented that demonstrate these claims.

The implications of these findings have both theoretical and practical importance. In its most basic form, the identification of processes related to time-space telescoping provides a counterpoint perspective to effects typically discussed for the EKC. In its simplest form, the EKC model suggests that all countries will 'grow out' of their environmental problems with increasing wealth, and that technological change keeps making this easier. The time-space telescoping perspective suggests that, in contrast, the different development milieu under which development now takes place can potentially exacerbate the scale and complexity of environmental challenges facing developing nations. Practically, it suggests that cities in the rapidly developing world need to reconsider environmental policies effectively used in industrialized countries. Their specific conditions necessitate more holistic and synergistic responses.

Notes

1   See, for example, http://www.resalliance.org/ and http://sustsci.harvard.edu/index.html.
2   Some have argued, on the other hand, that poverty contributes to environmental degradation, suggesting that economic growth is the answer to environmental challenges. This view, however, misses scale impacts. Indeed, the internal environments of poor neighborhoods and cities are dangerous and unhealthy places. These problems are typically relieved with increasing wealth but are replaced by other environmental challenges that impact the city as a whole. For further elaboration of this notion, see the section on environmental transition theory in this chapter.
3   Europe's average annual rates of population increase were highest between 1800 and 1900, as many countries in this part of the world were the first industrializers. During this period annual growth rates reached 1.0 per cent.
4   North America's fastest growth rates occurred between 1800 and 1850 when the country grew by 3.2 per cent annually. Growth slowed between 1850 and 1900 to 2.6 per cent, but this is still extremely rapid. This growth was not due to natural increases, but largely to immigration.
5   These include Brunei Darussalam, Cambodia, China, Hong Kong SAR, Democratic People's Republic of Korea, East Timor, Indonesia, Japan, Lao People's Democratic Republic, Macau, Malaysia, Mongolia, Myanmar, the Philippines, Singapore, Thailand and Vietnam.
6   Scholars now suggest that while population growth has been a significant driving force in environmental change over the past 200 years, the end of world population growth is now on the horizon (see, e.g., Lutz *et al.* 2004) Global population growth rate peaked at 2.1 per cent in the late 1960s, has since fallen to 1.35 per cent and is expected to continue falling.
7   Comparative GDP per capita incomes are provided in Geary-Khamis International dollars (see Maddison 2001).
8   The high aged population in China is related to the 'one-child policy'. Aging in China is described as 'aging first before affluence' (Miura 2000).
9   Variables were grouped into three elements including (1) 'financial' (i.e., capital interest, wages, bank deposits, etc.) (2) 'mixed character' (i.e., the volume and value of foreign trade, etc.); and (3) 'purely natural' (i.e., sector production and consumption levels).
10   In his in-depth analysis of long-term economic trends, Maddison (1991) finds no convincing evidence to support the notion of long waves, but does suggest, nevertheless, that there have been 'significant changes in the momentum of capitalist development. In the 170 years since 1820, one can identify separate phases which have meaningful internal coherence in spite of wide variations in individual country performance within each of them.' Maddison (1991) suggested that the move from one phase to another was governed by exogenous or accidental events that are not predictable.

11 A good example is the use of the automobile to solve problems associated with the use of horses for transportation and work related activities in cities (see, e.g., Melosi 2000).

12 This is not to say that changes in mining, timber extraction, agricultural (both species transfers and biotechnologies, and mechanical and chemical technologies) and water control technologies are not important. They are recognized as influencing environmental trends (for discussion of these see Headrick 1990).

13 These technologies include mobile and land line phones, satellite RVs, computer networks, electronic commerce, Internet services, etc.

PART III
TRANSFORMING STATE, SPACE AND SOCIETY: POLITICAL REFORM, PROMOTING CITIZENS RIGHTS AND POVERTY ALLEVIATION

Chapter 7

Civil Society Revisited: Travels in Latin America and China

John Friedmann

Over the last decade, the term 'civil society' seems to have become naturalized, at least in Anglo-American discourse. It is now used in everyday development talk, though mostly without a great deal of reflection. Non-government organizations are commonly referred to as 'civil society' or, alternatively, as the 'third sector' as an add-on to the sectors of state and private capital. Conservative commentators express concern over what they perceive as its dwindling energy in western countries. Typically, they link the recovery of civic energies to the process of democratization, not only in the new democracies of Eastern Europe but across the globe – the Middle East, for example, or the People's Republic of China. In their view, civil society is the last best hope of democracy; it is the people, acting through voluntary organizations in local communities who will take back government from corrupt politicians and impersonal bureaucrats and put it in the hands of associations they themselves control.

It is an attractive, if somewhat anarchistic vision of the future, a neo-liberal utopia. More progressive voices on the Left who have not yet given up on the social role of the state also champion civil society. But they see it less in terms of an ill-defined 'community' than of political action in progressive social movements, for example, against the war in Iraq, for the legalization of gay marriage, to save old-growth forests, to stop whaling, to protest globalization, for human rights and so forth in a never-ending series of struggles for a more humane world.[1] Rather than wishing to depoliticize the world, the Left sees an active civil society as the way to make governments more responsive to a shared desire for a socially and environmentally sustainable and peaceful world.

When 'civil society' can be used in such contrary ways indifferently by the Left and the Right, the term itself tends to lose meaning. Or at least it obliges us to critically reflect on its use. But rather than embark upon an archaeological expedition into political rhetoric with its constantly shifting meanings, a task for which others are better equipped than I, I propose to take you on a journey to South America where I first encountered the political rhetoric of 'civil society' in the 1970s, when it was indeed linked to a struggle for the return of democracy and a new discourse on citizen rights. In a brief theoretical interlude, I will then propose four models of civil society – the Tocquevillean, Habermasian, Castellsian and Gramscian – before continuing on our journey to China, or rather to the precincts of western universities where the question of a civil society in its possible relevance for

China has been intensively debated ever since the tragic debacle of Tiananmen Square in June 1989, when several hundred students were senselessly killed on orders of the Chinese Communist Party (CCP). I will then pose the question if any of the four models I have identified can be applied to China and with what consequences. In a brief conclusion, I will return to the underlying theme of democratization and the prospects of civil society.

Political Rhetoric of 'Civil Society' in Latin America

In Spanish Latin America, talking about the 'poor' is above all political talk. The very language of poverty tends to define the politics. Until quite recently, the subjects of poverty discourse were of course never asked to state their own views. The Catholic Church, however, which has an institutional concern for the 'poor', and conservative members of the political class are typically inclined to use the terminology of poverty with its implication of Christian piety. In a more secular vein, this same class refers to the 'poor' simply as *el pueblo*, in dim awareness that 'the people' are indeed a majority or near-majority in their country. By the time I arrived in Chile in 1965, the conservative governing coalition had just been replaced by left-of-center Christian Democrats, and the more common expression substituting for the somewhat dismissive *el pueblo* was now the 'popular sector' – *el sector popular*. In the development discourse of the time, the use of this term suggested that special programs would be launched to address some of the problems of poverty, such as affordable housing. A more activist language, championed by the Jesuits who advised the Christian Democrats on social policy, avoided the word poverty altogether, giving it a more sociological slant by speaking of *marginalidad* and *los marginados*, that is, of those who had been 'marginalized' – though it was never established who or what might ultimately be responsible for their marginalization – with the implication that the challenge was how to 'reintegrate' or 'reincorporate' the victims of marginalization with the mainstream of Chilean society.

During the 1970s, however, the discourse on poverty gradually took a more radical turn, especially in Brazil from where it spread throughout Hispanic America. Seemingly out of nowhere, political activists started to speak of 'citizens' (*ciudadanos/-as*), 'citizen rights', and *sociedad civil*. Two influences converged to produce this new language: Marxism and Liberation Theology, ideologies that were gaining prominence at the time. But whatever its inspiration, the meaning was clear. The 'poor' had entitlements by virtue of their citizenship, and it was the responsibility of the government to honor these entitlements in practice.

The frequent invocation of civil society occurred more or less simultaneously with a social movement called *basismo*, that is, a movement that concerned itself with the conditions of life of, say, the lower one-third of the social pyramid.[2] Its name derived from the idea of *comunidades de base* (base communities) or more accurately, ecclesiastical or Christian (base) communities, whose theological underpinnings were found in the writings of radicalized priests and theologians (Gutierrez 1973). Base communities, it was said, were modeled on early Christian gatherings in the Roman Empire. Usually convened by a local priest who ministered to the 'poor', they were ostensibly occasions for reading and interpreting the scriptures with a slant towards the social gospel. With diplomatic ambiguity, the Vatican referred to ecclesiastical communities as the 'popular church'. According to its advocates, the intention was to raise people's awareness of the 'objective' reasons for their abject condition of life and to instill in them a sense of their political rights as citizens.[3] Incidentally, in a time of authoritarian military regimes, base communities were also widely interpreted as a call for greater democracy.

Before long, the advocacy of liberation theology along with its real-world counterparts in the 'base communities' was squashed by a Vatican that feared the rise of the 'popular church' as a dissident movement and by the authoritarian governments that were flourishing with American support from the 1960s through the 1980s, most notably in Brazil, Argentina and Chile. The *basismo* movement had also to contend with competition from the Pentecostal and evangelical churches that were making headway throughout Latin America. What made Pentecostal Christianity so attractive to the political class was its non-political character. Among the popular sectors, its appeal was especially powerful among women, because it demanded abstinence from alcohol and promiscuity, provided for an emotional outlet in singing and fierce sermons and told believers that to get ahead in the world, one had to lead a disciplined, virtuous life in Christ and work hard, a Protestant version of the medieval monkish saying *ora et labora*. Then as now, the Pentecostal/evangelical survival strategy was strictly an individual one; it accepted the existing system of political and social relations as a given.

The popular appeal of Pentecostalism was one, if not the most important, reason why *basismo* ultimately declined (Burdick 1992). The missionary work of evangelical churches was well-financed through mother congregations in the United States, while progressive Catholics struggled to keep their 'popular church' filled. Not only was there an overall shortage of Catholic clergy in Latin America but of those who were already ordained, only a small minority was prepared to share the life of the poor. And those who elected to do so were regarded with suspicion not only by the hierarchy (the Vatican had reacted quickly by replacing 'red' Bishops with conservative stalwarts), but by the political class as well. The 'popular church' fostered feelings of (class) solidarity, and its strategy was ultimately a political one. But it offered no concrete practices through which its noble ends might be accomplished. Its structural account of misery was an abstraction that failed precisely where the 'moral rearmament' preached by the Pentecostals succeeded with their basic message: stop drinking, stop whoring, work hard. Properly motivated, anyone might be able to accomplish these mandates.

A secular version of *basismo* sprang up more or less simultaneously among some neo-Marxist intellectuals and others who came together as advocates of an alternative development. This intellectual movement had grown out of the dependency theory of underdevelopment – with its ostensible solution of an endogenous development. Among its leaders were economists, such as André Gunder Frank, Ignacy Sachs and Osvaldo Sunkel, all with long experience in Chile and Brazil, but the international headquarters of the movement was in a small town on Lake Geneva, headed by an idealistic Swiss national, Mark Nerfin. My own orientation followed this movement to the point of publishing a small book, *Empowerment*, subtitled *The Politics of Alternative Development* (Friedmann 1993). By then, however, the Swiss office had already closed its doors, and many of the arguments for an alternative development were being mainstreamed into World Bank policy, including a focus on poverty, the natural environment and the role of women.

In general, advocates of an alternative development had followed the *basismo* belief in community-based action that would leave the larger system of economic relations untouched. It was widely understood to be an approach to solving some everyday problems of poverty within a system that, left to itself, continued to generate poverty. During the past 15 years, or more, its banner has been taken up by so-called non-governmental organizations (NGOs) – some that are faith-based, others funded by international donor agencies, but increasingly also NGOs working under government contracts – that view themselves as its principal agents (Edwards and Hulme 1996).[4] Not surprisingly, NGOs and related community-based organizations are now generally identified as the 'third sector', that is, as 'civil society', and are working hand-in-hand with the governments of the day. Their work is characterized by attention to popular participation and self-help efforts by local communities of the poor, while the bulk of development aid goes towards infrastructure that meets the requirements of global capital. Although broadly speaking, their focus is on social projects, an overall anti-poverty strategy has not emerged, and NGOs do not, as a rule, succeed even in coordinating their efforts within local communities where several NGO-sponsored projects might compete for the attention of local residents. On the contrary, many of them act like business enterprises that see each other as competitive rather than complementary. And like the Pentecostals, most NGOs avoid taking a political stance and are prepared to work with the powers that be. Their so-called poverty alleviation strategy gives the political class free reign to pursue its own agenda.

It would be wrong, however, to say that nothing significant has happened in Latin America since the mid-1960s. Civil society has not been brought to heel. A powerful sense of citizenship and citizen rights has taken hold of disempowered groups and continues to be asserted. In Brazil, the Workers' Party (*Partido dos Trabalhadores –* PT), which is arguably an outgrowth of Paulo Freire's consciousness-raising work among rural workers in the 1960s (Freire 1972) and of the *basismo* movement in the 1970s and 1980s, has consolidated its mandate not only in key cities, such as Porto Alegre and Belo Horizonte, but reached the pinnacle of power in 2002 when it captured the country's Presidency under the leadership of Luiz Inácio 'Lula' da Silva. Brazil's political class now has to share power with a party that enjoys a strong popular base. According to Wright and Wolford (2003: 326), 'there has been a kind of explosion of citizen initiatives since the dying days of the dictatorship [in Brazil]'. Among the vigorous social movements they mention are a black consciousness movement with strong roots in local (urban) communities; the feminist and gay rights movements; a movement for indigenous rights; and an environmental movement. But the movement with the highest public profile is the militant *Movimento dos Trabalhadores Rurais Sem Terra* (MST – Landless Rural Workers Movement) that emerged in 1984 with the support of the left wing of the Catholic Church and the newly founded Brazilian Workers' Party PT. Today, with one million members and claiming to speak for five million dispossessed rural workers, it has adopted a strategy of land occupations – notably of large estates that are not being worked – and of establishing rural settlements called *agrovilas*, some of which have adopted cooperative (collective) forms of farming. These settlements offer a sense of community and, like the Pentecostals, enforce an ethic of virtuous living, that is, no alcohol, equal rights for women, hard work. They provide education and health services and are run along democratic lines (Rocha 2003; Wright and Wolford 2003).

Poverty in Brazil, as indeed elsewhere in Latin America, has of course not disappeared, and in some countries, it has probably deepened. But the disempowered masses are at least gaining a political voice and, conscious of their rights as citizens, are beginning to exert significant pressure on their governments for relief of their condition. Counter to early revolutionary dreams, they are fighting for inclusion in the system, not for its overthrow.

Four Models of Civil Society

In Latin America, civil society was a movement in a political struggle to reclaim citizen rights and a contested rhetoric about poverty and its conditions. But different actors at different times assigned different meanings to the term. Activists on the Left used 'civil society' as a fighting word to advocate democratization while claiming citizen rights for the poor. With the mainstreaming of an alternative development and the emergence of NGOs as the tool of a macro-strategy of poverty alleviation, civil society was renamed the 'third sector' and harnessed to hegemonic rule. The concept itself was never theorized.

A brief theoretical reflection on civil society will therefore be necessary before we continue our journey to the People's Republic of China. Without it, we cannot hope to make sense of how the concept is deployed by various scholars who, each in their own way, all hope for a democratization of Communist rule.

Over the past decade, a good many philosophical reflections on civil society have appeared in English (Cohen and Arato 1994; Keane 1998; Chambers and Kymlicka 2002). My intention here is not to review this literature; instead, I will propose four ways of 'reading' civil society. The conceptual setting for such a reading is taken from political economy, where civil society appears as a 'sphere of action' along with separate but overlapping spheres for state and (corporate) capital. To the extent that these spheres overlap, a public realm common to all macro-actors is created, a realm that, at least in principle, should be open to all citizens. In liberal democracies, this realm is identical with the polity as a whole. By implication, each of the three overlapping spheres of action is partly autonomous of the others, though in the nature of the case, this autonomy can never be absolute. The actions of our macro-actors will always be constrained by the existing strands of interdependency that link the three spheres to each other.

Given these broad theoretical considerations, the four models of civil society currently in use include the Toquevillean model of 'associative democracy', the Habermasian model of the 'public sphere', the Castellsian model of 'social movements', and the Gramscian model of 'hegemony'. It is to a brief sketch of each of these that I now turn.

Alexis de Tocqueville wrote his political analysis of *Democracy in America* in 1835 (Tocqueville 1988). Among the many things that impressed this aristocratic traveler from France about the freshly minted democracy he visited were the numerous community-based organizations throughout the small towns and cities of America, forming a dense pattern of associational life. From this emerged the model of an 'associative democracy', which posits 'intermediate' organizations between the state and the individual.[5] The model enjoys special favor today among neo-conservatives suspicious of the powers vested in the state. In their view, the best state is one that governs least and empowers local communities to deal with public issues as they arise through organizations such as local churches, benevolent associations, athletic clubs, neighborhood houses, and the like in which participation is voluntary. From time to time, these associations may become politicized around an issue of special interest to them, but for the most part, intermediate organizations of the Tocquevillean kind are active for themselves alone rather than in the public sphere.

The German political philosopher Jürgen Habermas formulated the second model of civil society in the course of his doctoral dissertation on the rise and decline of the 'public sphere' in Western Europe in the years following the French Revolution (Habermas 1989; orig. 1962). In his view, consistent with our conceptual setting above, the public sphere encompasses all matters that emerge as the 'common concerns' of a democratic polity. Its original model is the Athenian *agora* where democratic talk debated the issues of the day. In the late 18th century in France and England, a nascent bourgeoisie gradually became politically vocal finding its voice with the spread of newspapers, coffee houses, salons and a 'free-floating' intelligentsia. Drawing on this historical process, Habermasian civil society refers to people who in one way or another participate in democratic talk. We should remember, however, that democratic talk already assumes the existence of considerable freedom of speech and assembly, that is, of a certain form of political life. Where these conditions do not prevail, the Habermasian model cannot be said to apply.

Habermas himself talks about the rise and decline of the public sphere in Western Europe. In particular, he deplores its displacement in his own time by the manipulation of public opinion by both the political class and the state. There are numerous other problems with his concept of civil society, all of which, in one way or another, are a result of the vast imbalances in the distribution of effective power in society. It is thus an overly idealized, 'dialogic' model, though one that is perhaps, and for just that reason, useful as a template or norm against which to measure actual performance. In another, more radical interpretation, one could argue that rather than a model based on open dialogue, Habermasian civil society is an 'oppositional' or 'conflict model' in which a politicized *demos* brings pressure to bear on the political class for structural reforms to limit the accumulation of power in the hands of the few and to allow for greater and more effective participation by ordinary citizens in political discourse, in other words, for an inclusive public sphere.

The third model can be ascribed to the work of Manuel Castells and centers on the role of 'social movements'. In this perspective, social movements represent a mobilization of civil society for political ends (Castells 1983; 1997). Castells himself was born in Spain and educated in France; the major part of his academic career, however, was spent in the United States where he retired recently from the University of California at Berkeley.

It has become a commonplace to speak of the so-called 'new' social movements that became active in the decades following the end of World War II, among them the anti-nuclear movement, the anti-war movement, anti-globalization movement, the environmental and women's movements, the civil rights movement in the United States and any number of what Castells refers to as 'urban' movements for housing, gay rights, public transit, and so forth. Some of these social movements are essentially about resistance (anti-war, anti-nukes, anti-globalization); others are more pro-active and a source of socially progressive policy changes (indigenous rights, civil rights, women's rights). Social movements are characteristically self-organized and self-financed. They are non-violent in their actions – terrorism is therefore not a social movement of a mobilized civil society – and most of them espouse strong moral convictions. Their organization is around small face-to-face groups that are networked into larger structures as needed, although formal organizations, such as the World Wildlife Fund (WWF), sometimes carry on the more routine work of the movement. Leadership is therefore diffused, and movements often have the capacity to act on a global scale, commensurate with the influence of global corporations.

Our fourth model of civil society, the Gramscian, is based on the concept of 'hegemony'.[6] Antonio Gramsci, the Italian Communist theorist who languished for many years in a Fascist jail in Italy and died in 1935, asks an analytical question in his famous *Prison Notebooks* (Gramsci 1971): How do regimes of power maintain themselves with only minimal resort to overt coercion and violence? His answer is that this is accomplished through their ability to impose the hegemony of their ideology and culture by building a vast social consensus in support of the legitimacy of the existing system of power. This feat is accomplished through the public media, the educational apparatus, the entertainment industry, advertising and other means, all of which fall more or less into line in support of the dominant system of relations. Dissenting voices, even when they are permitted rather than quelled, are either marginalized or co-opted.

Given this definition, the strategic question for Gramsci was how to disrupt societal consensus and establish counter-hegemony in the midst of a consolidated hegemonic system. The Italian Communist Party seized upon this notion and adopted what was in effect a cultural strategy, by creating a complete society of the like-minded within the larger society of which it was a part, but which it sought to win over to its side. To guide this struggle, Gramsci formulated his famous doctrine of the 'war of position' (war in the trenches) and the 'war of movement'. In this context, civil society referred to members of the working class and lower bourgeoisie whose consensus needed to be disrupted by offering them the possibility of an ethically superior way of life – and its underlying ideology – to the one that they still considered to be the 'natural order' of things. To the extent that the Italian Communist Party was successful in this endeavor, Gramsci thought, it would gradually delegitimize the regime in power, and at some point, the war of position would transform itself into a war of movement in the course of which, deprived of its ideological support, the old regime would collapse and the Party would establish the new hegemony of a socialist society. In essence, the Gramscian model of civil society is a conflict model with a potential revolutionary outcome.

We are now ready to apply these four models to the Latin American case. In the first phase, guided by Liberation Theology and various interpretations of Marxist thought and, not least, informed by an idealistic reading of the Cuban revolution, activists on the Left toyed with what could be called a modified Gramscian version of civil society. Both Christian base communities (for the Catholic Left) and the communitarianism implicit in much of the writing on alternative development (including my own) set their hopes on the possibility of a counter-hegemony that would eventually challenge the status quo, ushering in not only a genuine democracy but a new 'social' economy that would embrace the least empowered sectors of civil society. These hopes were dashed. Put simply, an effective political agency to usher in the 'good society' was missing. More to the point, perhaps, communitarianism lacked broad public appeal even among the poor, though it still has a tenuous hold in the Brazilian landless movement whose cooperative settlements prefigure the socialist society its activists dream about.

In the second phase, when base communities were already history and many aspects of alternative development were being mainstreamed into foreign aid policy (e.g., by the World Bank), and with the rise of non-governmental organizations, the Tocquevillean model of civil society could be said to have prevailed. NGOs were re-baptized the 'third sector' and harnessed to a hegemonic policy of 'capitalism with a human face', the so-called poverty alleviation programs. Being non-political for the most part, and with their ambit of concern limited to local practice, the political threat NGOs might have posed to the state was deftly removed, and civil society was rendered harmless. It was now being touted as 'social capital' and thus assimilated to economics. As regards the Habermasian communicative model, it had no purchase anywhere, even though after the 1980s, more or less liberal democracies reappeared throughout the Luso-Hispanic world.

In the second version of a Habermasian civil society, in which the *demos* pressures the state for structural reforms, one could argue that to the extent that this was happening – as indeed it did happen in Brazil with the election of a Workers' Party government in 2002 and with the recent success of popular protests, spearheaded by militant labor unions and indigenous associations, in bringing down a traditional authoritarian regime in Bolivia – the conflict version of the Habermasian model may be applied. Today, however, the operative model is chiefly the Castellsian model centered on social movements. The *basismo* movement of the 1960s was an example of such movements, but its day has come and gone. Latin America now has other social mobilizations, among them movements for the rights of indigenous people, for women's rights, and for saving the environment.

The Prospect of Civil Society in China

We are now ready to shift our attention to China and the prospects there of a civil society. The tragic events on Tiananmen Square in 1989 set off a small cottage industry on the question of the prospective role of civil society in the democratization of the one-party state (Gold 1990; Wakeman 1993; Huang 1993; Gu 1993/4; Ding 1994; White *et al.* 1996; Brook and Frolic 1997; Chamberlain 1998; Pei 1998; Cheek 1998). The recent collapse of the communist states in

Eastern Europe, especially in Poland (with the role played by the Solidarity movement) and Czechoslovakia (thanks to the influence of writers like Václav Havel), had given rise to what can only be called a mythological account of the role of organized civil society in the regime changes that followed (Cohen and Arato 1994). A number of western scholars whose work focused on China had taken this into account, or were at least inspired by it. But the new discourse that resulted stood in stark contrast with Latin America, where, as we saw, the invocation of civil society occurred in the course of a long struggle for democratization and social justice. It was 'endogenous' to that struggle, whereas the discourse on China was largely the handiwork of western – chiefly Anglo-American – authors. Except for a small number of dissidents who had escaped to the West, Chinese intellectuals were not drawn into the debate about civil society.[7]

For the most part, proponents of the civil society thesis based their argument on Jürgen Habermas' doctoral dissertation on the rise and decline of the public sphere in Western Europe beginning in the late 18th century. The English translation of this careful historical study had just become available and offered an attractive foundation for what in any event was a speculative exercise (Habermas 1989). But now, 14 years after Tiananmen Square, it is safe to say that a Habermasian public sphere did not then nor does it exist now in China. Since 1949 and right on through the reform era, questions of common concern have been the exclusive domain of the Chinese Communist Party and therefore the domain of the state. A civil society 'beyond the reach of the state' and engaged in democratic talk about matters of common concern is simply unthinkable in today's China.

Not only is the public sphere preempted by the party-state, the basic institutional preconditions for its emergence still do not exist in China: freedom of speech and of assembly, and what we in the West like to speak of as the rule of law. Nor are these basic institutional conditions, which for us define a liberal democracy, likely to make their appearance in China any time soon. As Timothy Cheek has observed, 'The discourse of positive rights, limited government, and antagonistic public contention remains alien to Chinese intellectuals in general' (Cheek 1998: 251).

Similarly, we have to reject the Castellsian and Gramscian models. It was again Timothy Cheek (1998) who first introduced Gramsci's concept of hegemony into the civil society discourse about China. But he failed to spell out the complete model, which, as I have suggested, incorporates the idea of a counter-hegemony as well as a strategy for transforming the latter into the new hegemony of a victorious 'vanguard' party. It's hardly necessary to point out that a revolutionary model of this sort would be anathema to the Chinese Communist Party (CCP) that would bend all of its efforts to prevent a counter-hegemony from ever emerging. Equally unthinkable is the rise of a mobilized civil society in the form of social movements that are not tightly controlled by the CCP, as was the Great Proletarian Cultural Revolution instigated by Mao Zedong (a revolution within the Party *against* the Party). A repetition of that calamity is unlikely, but a genuine movement grounded in an autonomous civil society is similarly improbable.

This leaves us with the Tocquevillean model of 'intermediate' organizations – intermediate, that is, between state and individual. The number of organizations that might be called that has skyrocketed over the past 20 years, and their phenomenal growth has given rise to the thought that here, indeed, might be the beginnings of a 'third sector'. Specifically, the suggestion was made by Philip Huang (1993) in an essay for *Modern China*, a journal of which he was the founding editor. Huang argued that even though these 'civil' organizations were in many cases started by agencies of the state and continued to be supervised by the state, they might nevertheless, over time, claw back a substantial degree of independence from the state, evolving into a genuine civil society along Tocquevillean lines. Huang urged his colleagues to enter this new field of research. After a full decade, the first results of such research are in, and we can now critically assess Huang's claim.

In one of the best studies of Chinese 'intermediate' organizations, Gordon White (1996) develops a typology that distinguishes among four strata: a 'caged stratum' of mass organizations (such as the All-China Federation of Trade Unions); an 'incorporated stratum' of officially recognized social organizations (business, trade, professional, academic, sports, recreational and cultural), of which by late 1993 there were already 1460 at the national, 19,600 at the provincial, and 160,000 at the county level; an 'interstitial stratum' of mostly local organizations that while officially registered, and despite some degree of state supervision, are more or less free to pursue their interests without direct interference from above; and finally, a 'suppressed stratum' that includes a wide variety of political and social organizations as well as secret societies and other criminal organizations (White 1996: 196–222; see also the more thoroughly documented study by White *et al.* 1996).

Leaving aside the 'captive stratum', the question remains whether White's incorporated and interstitial strata could evolve into a version of civil society 'for itself' or even more hopefully, into a civil society capable of exercising the right to free speech and using this right to contest the state. The answer comes in part from an excellent study by Kenneth Foster (2002) of business associations in Yantai, a medium-sized city in Shandong province. He writes:

> [T]his paper argues that China's business associations (as well as many professional associations) can be more fruitfully studied as new elements of the state's administrative system than as participants in a state-society dialogue. That is, if we are to understand the emergence and role of business associations in China, we must first analyse how they are connected to state and Party organizations and how these links affect their character and operation. My research in the city of Yantai shows that nearly all business associations there were created at the initiative of state and Party officials, and that they are in essence appendages of government or Party organizations. Over the past two decades, these associations have accordingly functioned more as new parts of the local administrative system than as new sites of state-society engagement. However, this does not mean that they should be regarded simply as newer versions of the classic Leninist transmission belts. Unlike instruments of top-down control, business associations in Yantai are used by government agencies and their officials in a localized and parochial fashion to assist the agencies in carrying out their mandated tasks. But sometimes they are not used at all and become mere hollow shells (Foster 2002: 42).

Foster's picture of so-called intermediate organizations in Yantai does not bode well for a civil society, at least for now.[8] Even local Chambers of Commerce, it turns out, are frequently created by and integrated with the Industry and Commerce Federation, a 'tame organization under the Party's United Front Department'

(Foster 2002: 49). Funding for these organizations comes for the most part directly from the state or state-operated enterprises, although membership dues are sometimes collected, though not without difficulty.

There remains the possibility of the interstitial organizations mentioned by White. In a recent field report from Guanting Township in China's northwestern and impoverished Qinghai Province, the authors examine the case of a single rural NGO, the Sanchuan Development Association (SDA) (Zhang and Baum 2004). The SDA, they write, 'is a relatively autonomous, indigenous NGO of a type that appears to be popping up with increasing frequency in rural China, but that to date has not been widely reported or taken account of in the civil society/corporatism literature' (op. cit.: 99). SDA is entirely financed by foreign donors and is exclusively devoted to local poverty alleviation and community development projects, ranging from school construction, public sanitation and water conservation to horticulture and animal husbandry. The authors call it a 'true minjian' (people's) NGO. In the development literature, it would probably be referred to as a CBO (community-based organization). They observe that an exploratory Google.com search, using the keywords 'China+rural+poverty alleviation+international NGOs' generated almost 500 hits. As a foreign-financed CBO, the Sanchuan Development Association tries to keep the local state at arm's length – though it is registered and ostensibly 'supervised' by the state – and remains studiously non-political. It may count as an 'intermediate' organization, but it has little to do with transforming an autocratic into a more democratic regime.

If none of the models of civil society have any purchase in China, the prospects for democracy arising out of a civil society in that country are indeed dismal. The question is left why this should be the case. The disposition of the CCP to autocratic rule is not the full answer, and we may have to look further into China's cultural heritage, most notably the Confucian tradition that is endogenous to the region. Confucianism is a powerful tradition that has shaped local societies throughout northeast Asia for more than 2000 years, just as Christianity has done in the West.

Following the French revolution, civil society was forged as a concept in the Judeo-Christian tradition in which Jesus of Nazareth could say: 'Render under Caesar what is Caesar's and unto the Lord what is the Lord's.' The Lord's voice in this instance was directed straight at the mind and heart – the soul – of the single, stand-alone individual. It was this individual, in all of his/her ignorance and weakness, whose concerns needed the intermediation of an organized civil society *vis-à-vis* the state just as a priesthood was needed to mediate relations between God and Man. This ancient dichotomy of the ruler and the ruled whose interests were not necessarily identical and indeed were often at odds with each other required a civil society and political institutions that were designed to mediate conflict. It also gave rise to a dichotomous distinction of what is public and private, a distinction that would create considerable difficulties in reconciling the two realms.

If the foregoing is a roughly accurate account of the way those of us raised in the Judeo-Christian tradition tend to understand the world, the Confucian way offers another conception of public and private realms. Here, individuals are always already embedded in 'affective networks of social relation' whose paradigmatic form is the family and the extended kinship network of the clan. Korean scholars Lew, Chang, and Kim (2003: 214–15) provide a succinct statement of this tradition.

Zhu Xi [important neo-Confucian philosopher at the end of the 13th century whose writings led to the enthusiastic adoption of neo-Confucian practices in Korea] had made clear and elaborate provisions, both theoretical as well as institutional, for the reconstruction and strengthening of the institution of the family. The clan, or the family writ large, was the institution of choice for Zhu Xi just as it had been for Confucius, because it was where filial piety, loyalty, trust and other values essential for the affective society idealized by Confucius could flourish. As a sociopolitical institution, it provided the sense of continuity, permanence, and identity that the highly bureaucratized, impersonal, and commercialized society of imperial Song could not.

Home was therefore not merely the cornerstone of the private sphere. It was, at the same time, a public sphere 'in which one is taught one's rights and duties, responsibility, and power. Confucians viewed public and private spheres to be in harmony rather than in conflict'.

As northeast Asia's most Confucian country, South Korea provides us with insights into a society that, like China's, understands individuals first and foremost as beings whose very nature is defined by their social ties. This way of understanding who we are as human beings is not merely a 13th century conceit to be displaced by a more 'modern' idea of the individual but continues right up to the present era. According to Lew *et al.* (2003: 203–04):

Affective networks in Korea are based upon three factors: blood ties, school ties, and regional ties. Korean businesses are (in-)famous for the way in which they are family based. Most of the largest *chaebols* (conglomerates), including those most internationally competitive ... are still controlled by members of the founder's family, usually brothers, sons, nephews, and grandsons. The importance of blood ties is even greater for smaller companies ...

And the authors continue:

Given the importance of affective networks, it is little wonder that Koreans invest significant amounts to time and energy attending innumerable social gatherings such as weddings, funerals, and alumni meetings, as well as more traditional social groups ... This is in sharp contrast to the conspicuous lack of commitment to and involvement in the voluntary citizens' groups or 'civic organizations' deemed essential for a thriving democracy [i.e., the Tocquevillean model]. Although many Koreans would gladly pay US$10 for a round of drinks for 'old buddies', few are willing to pay even half that amount in fees and dues to citizen groups. Even in the smallest localities, town elite gatherings function as the focal points of affective networks linked by blood, school, and regional ties (Lew *et al.* 2003: 205).

Perhaps we can now begin to understand why so-called intermediate organizations in Yantai and presumably elsewhere in China are more often hollow forms than the substance of a vital civil society, why their members are disinclined to contribute to their support, and why the state nevertheless continues to create them to carry out its mission.[9] In China, as in South Korea, the private and public spheres have always been intertwined; they are not inherently in conflict. As the authors conclude,

What we have then, in Confucianism, is a philosophy of the public that does not follow or respect those dichotomies essential for the concept in the Western tradition, namely, 'individual versus the group, 'state versus civil society', and 'public versus private sectors.' ... [This] does not mean that it lacks a sense of the public as opposed to the private. Confucians are able to make the distinction between 'cronyism' and a 'just' order. It is just that they are not neatly distinguished along the dichotomous lines so dear to Western [Anglo-American?] political discourse (Lew *et al.* 2003: 215).

Conclusion

The position I have tried to argue in this chapter is that civil society is a concept embedded in liberal democracy where a term that would refer to the *demos* as an actor in the public sphere was a necessary complement to the more formal democratic institutions. Civil society was that actor, and it had to be sufficiently distinguished from the state – and later also from corporate capital – to endow it with a substantial degree of autonomy. It was thus both a social and a political concept. As the former, it included those organizations outside state and capital that were concerned with civic life. This was the Tocquevillean idea of an associative democracy. As a political concept, formulated in very different ways by Habermas, Gramsci, and Castells, it was supposed to engage the state through social movements and political parties. This was the arena of democratic talk. Essential to the life of civil society were democratic freedoms, most importantly the freedom of speech and association, and a legal system capable of safeguarding these freedoms against the state.

In actual use, civil society has more often than not been deployed as a rhetorical weapon in the struggle for democracy and social justice. The example from Latin America has demonstrated this use over the course of several decades in a struggle that eventually led to the consolidation of liberal democracy throughout Latin America and the emergence of non-governmental joined to community-based organizations in the continuing fight against disempowerment and poverty. The alternative development discourse that grew up alongside this struggle has meanwhile been mainstreamed into development programs, notably by the World Bank.

Civil society discourse about China turns out to be an altogether different matter. It is not a discourse endogenous to China; it is rather a discourse among western scholars who speculate about the relevance of the concept and China's prospects for democratization. These debates have raised new questions and produced new insights, but they are only tangentially related to whatever forces may lead to a democratic opening in China. Our review of the four most common models of civil society has, I believe, unambiguously demonstrated that these western models are inapplicable to the People's Republic of China.

I have argued that the basis for this judgment can be found in the very different intellectual/religious heritage of China – and more generally, Northeast Asia – as compared to the West. Whereas the Confucian tradition sees a vertically integrated society intertwined with the institutions of the state, and the private is subsumed under the public and thus has no independent status, the Judeo-Christian way sees only free-standing individuals in need of mediating social institutions between themselves and the powerful structures of the state.

We call these mediating institutions civil society, and they are found in varying measure only in countries that are heir to these traditions.[10] China, with a very different take on what it means to be human, has no need for a western-style civil society. What this implies for the prospects of democracy in China remains an open question. But whatever form it takes, it is unlikely to resemble, except in the most superficial sense, a liberal democracy in our (western) common understanding.[11]

Acknowledgements

I would like thank Jacqueline Chase for reading and commenting on the Latin American section of this chapter, Kenneth Foster for commenting on the section dealing with China, and Timothy Cheek, both for his suggestion that I take a comparative approach to the civil society debate about China and for his critical comments on an earlier version of this chapter.

Notes

1 There are of course also conservative or reactionary social movements that could with equal force argue their civil society status.
2 For an account of *basismo*, its origins and spread, see Levine and Mainwaring (1989) and Lehman (1990).
3 A major influence here was Paulo Freire. See his famous Pedagogy of the Oppressed (Freire 1972).
4 The number of NGOs has mushroomed in recent years. One estimate puts their numbers in Brazil at 210 000. In India, there are said to be roughly one million. But detailed case studies of NGOs are surprisingly lacking.
5 Note the formulation of this model, which has the individual as the irreducible social atom rather than a collective entity, whether it be the family, the clan, or other social formation (e.g., 'community').
6 I am grateful to Timothy Cheek for pointing out what for me was an unfamiliar take on civil society. See Cheek (1998).
7 According to Kenneth Foster, Chinese intellectuals have begun to talk about civil society since the beginning of the new millennium (Personal communication).
8 The point is put more strongly by Lew *et al.* (2003: 206) even though their country of reference is a democratic South Korea: '… a genuine civil society outside and independent of the state does not exist in Korean society'. In East Asia, democracy takes other forms than the liberal state we cherish in the West.
9 In an interesting article, Yang (2003) argues that China is experiencing what he calls a co-evolution of the Internet and civil society, that is, the beginnings of a 'virtual civil society' that manages to exist within the officially set parameters of political discourse. Yang's findings are based on two case studies of web-based bulletin boards and chat rooms. If the idea of co-evolution is correct, we can expect a vast expansion of a virtual civil society in China as Internet penetration reaches further and further down the economic ladder. But will it be introvert (for itself) or extrovert (aimed at claiming and defending liberties and other rights)?
10 Chinese-style familism is more prevalent in countries such as Italy, Spain, and Portugal, as well as Argentina and Chile, and civil societies in these countries also tend to be

relatively weak. It may be noted that it is precisely these countries that have had the most difficulty, though not insuperable, in living with liberal democracy.

11 Nominally liberal democracies exist in South Korea, Japan, and Taiwan. But studies comparing these political systems to those prevailing in North America or Western Europe would undoubtedly show major differences. For the beginnings of a research agenda along these lines, see Chang (2003) and Hall and Ames (2003).

Chapter 8

State and Urban Space in Brazil: From Modernist Planning to Democratic Interventions

Teresa Caldeira and James Holston

In the last half century, the Brazilian state consolidated and then liquidated a modernist model for the production of urban space. According to this model, best crystallized in the construction of Brasília, the state produces urban space according to centralized master plans that are conceived as instruments of social change and economic development. The role of government is both to articulate these plans and to create the means for their realization. During the last two decades, however, a constellation of forces – including main elements of the state, business and industry, popular social movements, political parties, and non-governmental organizations (NGOs) – rejected this centralized conception of state intervention. In its place, they substituted a notion of planning in which government does not produce space directly, but rather acts as a manager of localized and often private interests in the cityscape. Moreover, whereas the modernist model entails a concept of total design, by which planners impose solutions, like demigods, the new model considers that plans should both be based on and foster the exercise of democratic citizenship.

The new planning results from a confluence of contradictory factors. On the one hand, many of its tenets were first proposed by social movements and NGOs concerned with urban reform in the 1970s and 1980s. Some of the most significant of these principles were included in the new Federal Constitution of 1988 (called the Citizens' Constitution) and were developed in subsequent urban plans and legislation by these organizations. Therefore, the new model of planning is an explicit expression of the democratization process that has been transforming Brazilian society and its ways of conceiving citizenship since the 1970s. On the other hand, the same instruments have also been used by some municipal administrations and powerful private organizations to produce the contrary of their original intent, namely, the privatization of public space, spatial segregation, social inequality, and private real estate gain. Moreover, the redefinition of the role of the state expressed in the new planning cannot be associated with democratization alone. In addition, the collapse of the interventionist modernist mode relates to a fiscal crisis of the state, industrial restructuring, and the adoption of neoliberal policies usually justified as necessary to keep Brazil in pace with the new demands of globalization.

This chapter contrasts these two models of governmental production and management of urban space. It also addresses the consequences of each for the lives and spaces of the working-class people who inhabit both the poor peripheries of Brasília and São Paulo and the *favelas* and *cortiços* of their centers.[1] Since the beginning of industrialization, governmental production of space in Brazilian cities has meant the creation of a legal and regulated city for the upper classes and an illegal and unregulated city for the majority of the working poor; that is to say, for the vast majority of Brazilians. Illegality and improvisation have always been the conditions under which the urban poor have created their spaces in Brazilian (and most Third World) cities. The instruments of urban policy created during the democratic period attempt to address the problem of illegality and therefore of social injustice in Brazilian cities. Nevertheless, as powerful corporations and real state interests engage these same instruments, they generate new forms of spatial segregation and undermine some of the paths to urban improvement and citizenship expansion that the social movements of the 1970s and 1980s achieved.

Total Planning

Owing to the need to constitute a base of radiation of a pioneering system [of development] that would bring to civilization an unrevealed universe, [Brasília] had to be, perforce, a metropolis with different characteristics that would ignore the contemporary reality and would be turned, with all of its constitutive elements, toward the future.
President Juscelino Kubitschek 1975: 62–63

The apartment blocks of a superquadra [Brasília's basic residential unit] are all equal: same façade, same height, same facilities ... which prevents the hateful differentiation of social classes; that is, all the families share the same life together, the upper-echelon public functionary, the middle, and the lower.
Brasília 1963 [65–81]: 15

As exemplified by Brasília, total planning in Brazil cannot be separated from either modernism or developmentalism.[2] Even before the construction of Brasília, the Brazilian government had appropriated the international model of modernist architecture and planning developed by CIAM (*Congrès Internationaux d'Architecture Moderne*). Its intention was to use this model to create a radically new urban development as a means to overcome the nation's backwardness, as a means to bring the nation, through leaps in history, into the vanguard of modernity. Modernist total planning is an instrument of social transformation as much as of spatial production. It is conceived as a means of creating an urban environment that molds society in its image. This two-fold transformation brings progress and development. Brasília is no doubt the most complete example ever constructed of the CIAM model city – a model that dominated urban theory and policy in many countries for most of the 20th century, from the 'new cities' of Eastern Europe to the 'edge cities' of American suburbia. In Brazil, this conception of planning reigned supreme from the 1940s to the 1980s. As Brazil became highly urbanized and industrialized during this period, it shaped most of the state's urban and economic undertakings.[3]

Both Brasília and modern São Paulo took form under the influence of a nationalist ideology of modernization known as 'developmentalism'. Briefly, the idea was to use direct state intervention to promote, in a concentrated period of time, national industrialization based on import substitution. Its main objective was to produce not only accelerated industrialization but also modern subjects, that is, rational and 'domesticated' consumers for its products. The slogan of President Juscelino Kubitscheck's Target Plan of development in the mid-1950s was '50 years in five'. This model of development sustained not only São Paulo's industrialization but also the construction of Brasília and other state-sponsored projects aimed at turning Brazil into a modern nation.[4] To promote progress through leaps in history, the Brazilian state took upon itself a wide range of tasks that included building cities, roads, and electric plants, sponsoring industrial production (especially of automobiles, chemicals, and steel), as well as expanding the welfare state and modernizing television programs. From factories to hospital networks, from mines to television stations, from telephone companies to universities, all materialized under the control and usually the ownership of the state.

Shared by citizens of all social classes, a strong faith in progress anchored the developmentalist project of the Brazilian state.[5] From the 1950s to the 1980s, Brazilians believed massively that Brazil was *o país do futuro* (the country of the future). Especially in the major cities, people supposed that hard work would bring individual betterment, modern urbanization to the urban peripheries where most lived, and general prosperity through industrial expansion. The sum of these achievements would produce the modernization of Brazil. Although it soon became clear that modernization would not significantly reduce the enormous inequalities separating rich and poor, Brazilians continued to believe that progress would nevertheless benefit all.

Brasília was the most accomplished symbol of this project of progress, development and modernization. Its founders envisioned Brasília's modernist design and construction as the means to create a new age by transforming Brazilian society. They saw it as the means to invent a new nation for a new capital – a new nation to which this radically different city would then 'logically belong' as its planner Lucio Costa claimed (1980: 15). This project of transformation redefines Brazilian society according to the assumptions of a particular narrative of the modern, that of the CIAM modernist city, most clearly expressed in Costa's Master Plan and in the architecture of Oscar Niemeyer, the city's principal architect.

As universally acknowledged, the project of Brasília is a blueprint-perfect embodiment of the CIAM model city. Moreover, its design is a brilliant reproduction of Le Corbusier's version of that model.[6] Nevertheless, Brasília is not merely a copy. Rather, as a Brazilian rendition of CIAM's global modernism, its copy is generative and original. Brasília is a CIAM city inserted into what were the margins of modernity in the 1950s, inserted into the modernist ambitions of a post-colony. In this context, the very purpose of the project was to capture the spirit of the modern by means of its likeness, its copy. It is this homeopathic relation to the model, brilliantly executed to be sure, that gives the copy its transformative power. In other words, its power resides precisely in the display of likeness. This display of an 'original copy' gives the state a theatrical form, a means to construct itself by putting on spectacular public works.

As the exemplar of this stagecraft-as-statecraft, Brasília was designed to be a mirror for the rest of Brazil, reflecting the modern nation Brazil would become. It was conceived as a civilizing agent, the missionary of a new sense of national space, time and purpose, colonizing the whole into which it has been inserted. To build the city in just three and a half years, Novacap, the company in charge of the construction, instituted a regime of round-the-clock construction. This regime of hard work became known throughout Brazil as the 'rhythm of Brasília'. Breaking with the meters of colonialism and underdevelopment, this was a new rhythm, defined as 36 hours of nation-building a day – '12 during daylight, 12 at night, and 12 for enthusiasm'. It expresses precisely the new space-time consciousness of Brasília's modernity, one that posits the possibility of accelerating time and of propelling Brazil into a radiant future.

The rhythm of Brasília thus reveals the development of a new kind of agency, confident that it can change the course of history through willful intervention, abbreviating the path to the future by skipping over undesired stages of development. This modernist agency of rupture and innovation expressed itself in all domains of Brasília's construction and organization, from architecture and planning to schools, hospitals, traffic systems, residential organization, property distribution, bureaucratic administration, music, theater, and more. Brasília's modernism signified Brazil's emergence as a modern nation because it simultaneously broke with the colonial legacies of underdevelopment as it posited a radiant future of industrial modernity. The new architecture and planning attacked the styles of the past that constituted especially visible symbols of a legacy the government sought to supersede. It privileged the automobile and the aesthetic of speed at a time when Brazil was industrializing. It required centralized planning and the exercise of state power that appealed to political elites.

To create a new kind of society, Brasília redefines what its Master Plan calls the 'key functions of urban life', namely work, residence, recreation and traffic. It directs this redefinition according to the tenets of the CIAM model city. CIAM manifestos call for national states to assert the priority of collective interests over private. They promote state planning over what they call the 'ruthless rule of capitalism', by imposing on the chaos of existing cities a new type of urbanism based on CIAM master plans. CIAM's overarching strategy for change is totalization: its model city imposes a totality of new urban conditions that dissolves any conflict between the imagined new society and the existing one in the imposed coherence of total order.

One of the principal ways by which CIAM design achieves its totalization of city life is to organize the entire cityscape in terms of a new kind of spatial logic. As we have analyzed this logic elsewhere (Holston 1989), we do not pause to examine it here except to say that its subversive strategies have overwhelming consequences for urbanism, especially its elimination of the corridor street and related public spaces and its inversion of Baroque solid-void/figure-ground relations. Complementing its theory of spatial change, the CIAM model also proposes a subjective appropriation of the new social order inherent in its plans. It utilizes avant-garde techniques of shock to force this subjective transformation, emphasizing decontextualization, defamiliarization and dehistoricization. Their central premise is that the new architecture/urban design creates set pieces of radically different experience that destabilize, subvert, and then regenerate the surrounding fabric of social life. It is a viral notion of revolution, a theory of decontextualization in which the radical qualities of something totally out of context infect and colonize that which surrounds it with new forms of social experience, collective association, personal habit, and perception. At the same time, this colonization is supposed to preclude those forms deemed undesirable by negating previous social and architectural expectations about urban life.

Brasília's design implements these premises of transformation by both architectural and social means. On the one hand, its Master Plan displaces institutions traditionally centered in a private sphere of social life to a new state-sponsored public sphere of residence and work. One of its most radical tenets in this regard was the elimination of private property altogether in favor of state ownership – at least until 1965, when the military government created a private real estate market. On the other hand, Brasília's new architecture renders illegible the 'taken-for-granted' representation of social institutions, as the buildings of work and residence receive similar massing, siting and fenestration, and thereby lose their traditional symbolic differentiation.

No one should doubt the potency of these modernist strategies of defamiliarization. In Brasília, they proved to be brutally effective, as most people who moved there experienced them with trauma. In fact, the first generation of inhabitants coined a special expression for this shock of total design, *brasilite* or 'Brasília-itis'. As one resident told Holston, 'Everything in Brasília was different. It was a shock, an illusion, because you didn't understand where people lived, or shopped, or worked, or socialized.' Another common disorientation is the sense of exposure residents experience inside the transparent glass façades of their apartments. Thus, Brasília's modernism also works its intended subversion at an intimate scale of daily life. Harmonized in plan and elevation, Brasília's total design created a radically new world, giving it a form that possessed its own agenda of social change.

In summary, modernist master planning as exemplified in Brasília is a comprehensive approach to restructuring urban life precisely because it advances proposals aimed at both the public and the private domains of society. Its proposals for the former focus on eliminating the street and its public aspects, both spatial and social. Its proposals for the latter center on a new type of domestic architecture and 'collective' residential units. Its design restructures the residential not only by eliminating private property but also by reducing the social spaces of the private apartment in favor of a new type of residential collectivity in which the role of the private and the individual is symbolically minimized – by using transparent glass façades, eliminating traditional informal spaces, and so forth.[7] Together, these strategies constitute a profound estrangement of previous modes of urban life, achieving a similar kind of defamiliarization of public and private values in both the civic and the residential realms.

It is important to emphasize that the CIAM modernist model is strongly egalitarian in motivation. As the epigraph of this section indicates, its objective is to impose the means of equalization 'to prevent the hateful differentiation of social classes'. Hence, it develops a new type of urban environment both to eliminate previous expressions and instruments of inequality and to force people to behave in new ways that the planners envision ('the same life together'). The model's commitment to equalization is remarkably comprehensive, aimed at transforming both public and intimate relations of social life. Although committed to equalization, however, modernist planning is decidedly not democratic. Rather, it is based on an imperial imposition of its brand of panoptic equality, a 'planner knows best' vision of an already scripted future. As we shall see, implemented as a means to mediate equality, it fails perversely.

The radically new world of Brasília immediately confronted a classic utopian dilemma, one inherent in all forms of modernist planning: the necessity of having to use what exists to achieve what is imagined destroys the utopian difference between the two that is the project's premise. As Brasília's demigods – the planners – struggled to keep pace with the vitality of the city they had brought to life, their directives revealed two fundamental features of the modernist mode of governmentality: first, they maintained the priorities of the plan at all costs, not admitting any compromise with 'what exists', with contingent developments or history's engagement with the ideal. Second, their reiterations of the plan to counter contingency turned the project of Brasília into an exaggerated version of what the planners intended to preclude. In effect, they reproduced the Brazil they wanted to exclude. This Brazilianization contradicted many of the Master Plan's most important intentions.

One of the clearest examples of this perversion is the reproduction in the new capital of a legal center and an illegal periphery. The government planned to recruit a labor force to build the capital, but to deny it residential rights in the city it built for civil servants transferred from Rio de Janeiro. By 1958, however, it became clear that many workers intended to remain. In fact, almost 30 per cent of them had already rebelled against their planned exclusion by becoming squatters in illegal settlements. Yet the government did not incorporate the *candangos* (the pioneering construction workers) into the *Plano Piloto* (as the modernist city itself is called), even though it was nearly empty at inauguration. The government found this solution unacceptable because inclusion would have violated the preconceived model that Brasília's 'essential purpose [was to be] an administrative city with an absolute predominance of the interests of public servants' (Ministry of Justice 1959: 9). Rather, under mounting pressure of a *candango* rebellion, and in contradiction of the Master Plan, the administration decided to create legal satellite cities, in which *candangos* of modest means would have the right to acquire lots and to which Novacap would remove all squatters. In authorizing the creation of these satellite cities, the government was in each case giving legal foundation to what had in fact already been usurped, namely, the initially denied residential rights *candangos* appropriated by forming illegal squatter settlements. Thus, Brasília's legal periphery has a subversive origin in land seizures and contingency planning.

Modernist planning attempts to overcome the contingency of experience by totalizing it, that is, by fixing the present as a totally conceived plan based on an imagined future. Holston (2001) contrasts this model with what he calls 'contingency planning'. The project of Brasília generated both modes. Although both were experimental and innovative at the time, they were – and remain – fundamentally at odds. Contingency planning improvises and experiments as a means of dealing with the uncertainty of present conditions. It works with plans that are always incomplete. Its means are suggested by present possibilities for an alternative future, not by an imagined and already scripted one. It is a mode of design based on imperfect knowledge, incomplete control and lack of resources, which incorporates ongoing conflict and contradiction as constitutive elements. In this sense, it has a significant insurgent aspect, though it may have a regressive outcome. The built Brasília resulted from the interaction of both modes of planning: the total and the contingent. In most cases, however, the former soon overwhelmed the latter in the development of the city.

For example, to remain faithful to their modernist model, planners could not let the legal periphery of satellite cities develop autonomously. They had to counter contingency, in other words, by organizing the periphery on the governing rationality of the center. To do so, they adopted what we can call a strategy of retotalization, especially with regard to the periphery's urban planning, political-administrative structure, and recruitment of settlers. This strategy had two principal objectives: to keep civil servants in the center and others in the periphery, and to maintain a 'climate of tranquility' that eliminated the turbulence of political mobilization (Ministry of Justice 1959: 9). Given these objectives, the planners had little choice but to use the mechanisms of social stratification and repression that are constitutive of the rest of Brazil they sought to exclude. First, they devised a recruitment policy that preselected who would go either to the center (*Plano Piloto*) or to the periphery (satellite cities) and that would give bureaucrats preferential access to the former. Second, in organizing administrative relations between center and periphery, planners denied the satellite cities political representation. Through this combination of political subordination and preferential recruitment, of disenfranchisement and disprivilege, planners created a dual social order that was both legally and spatially segregated. Ironically, it was this stratification and repression and not the illegal actions of the squatters that more profoundly 'brazilianized' Brasília.[8]

Predictably, the reiteration of the orders of the center in the periphery created similar housing problems there. These problems led, inevitably, to new land seizures and to the formation of new illegal peripheries – now in the plural because each satellite spawned its own fringe of illegal settlements. Moreover, by the same processes, some of these seizures became legalized, leading to the creation of yet additional satellite cities. These cycles of rebellion and legitimation, illegal action and legalization, contingency planning and retotalization, continue to this day. A striking illustration of the perpetuation of Brasília's contradictory development is that, even today, the *Plano Piloto* remains more than half empty while only containing 13 per cent of the Federal District's total population. This comparison strongly suggests that the government continues to expand the legal periphery rather than incorporate poor migrants into the Plano Piloto.[9] As a result, Brasília remains Brazil's most segregated city (Telles 1995).

Modernization without Substantive Citizenship

Most other Brazilian metropolitan regions have not been the product of such direct and total planning as Brasília. Nevertheless, the oppositions between legal and illegal urban areas, center and periphery, and rich and poor are equally constitutive. This is the case of São Paulo, a city that has also come to symbolize Brazil's modernity by concentrating the largest share of the country's industrial production, economic growth and urbanization. São Paulo's decisive turn to industrialization dates from the 1950s and shares some of the same instruments and visions of Brasília, including the use of modernist design and the notion that the city had to be opened up for circulation.[10] The new industries were placed outside the center. As industrialization intensified and migration reached its peak in the 1950s, the local administration was busily opening avenues and removing the remaining tenement housing downtown. The modern city that emerged was disperse and organized by clear class divisions. The center received improvements in infrastructure and the most obvious symbols of modernity: it was dominated by skyscrapers – increasingly of modernist design – that multiplied in a matter of a few years from the 1950s on, and gave the city its contemporary identity.

In the periphery, the rhythm of construction was no less intense than in the center. But the lack of any kind of state support, investment and planning generated a very different type of space. On the outskirts of the city, workers bought cheap lots of land sold either illegally by outright swindlers or with some kind of irregularity by developers who failed to follow city regulations regarding infrastructure and land registration. In spite of their illegal or irregular activities, these developers received a free hand from successive generations of city administrators, who preferred to close their eyes to what was happening in the periphery and to administer only the 'legal city'.[11] As for the workers of Brazil – in São Paulo, Brasília, and elsewhere – they have always understood that illegality was the condition under which they could have access to land and inhabit the modern city. To them, residential illegality signifies not just material precariousness and distance from the center, but also the possibility of becoming modern and of establishing a claim to eventual property ownership. In streets without pavement and infrastructure, workers built their own houses by themselves and without financing. This could only happen through a slow and long-term process of transformation known as *autoconstrução* (autoconstruction) (see Holston 1991a). It is also a process that perfectly represents progress, growth and social mobility: step-by-step, day-after-day, the house is improved and people are reassured that sacrifice and hard work pay off. Thus, workers moved to the 'bush' to build their houses and, through the process of *autoconstrução*, were the agents of the peripheral urbanization of the city. That the population density of the city decreased by half between the beginning of the 20th century and the 1960s, in spite of remarkable population growth, indicates the enormity of this expansion.[12] As a result, the urbanized area of the city of São Paulo more than tripled between 1930 and 1954, and doubled again by the 1990s to reach its current size of 850 sqkm.

Thus, in both São Paulo and Brasília, governmental strategies toward modernization, industrialization, urbanization and development were sometimes interventionist and other times *laissez-faire*. However, they resulted in a similar structure of urban inequality. In both cities, these strategies reveal an overarching conception of how to govern society and produce its modernity. The general principle is to govern without generating social equality or turning the masses into active citizens. The split between legal and illegal symbolizes succinctly the underlying perspective of Brazilian elites on modernization: those considered non-modern (i.e., the vast majority of the population) were incorporated into their plans as a labor force but marginalized as citizens. They were denied the right to vote, excluded from legal property in the modern cities, and violently silenced by the military dictatorship.[13]

Although developmentalist-modernist planning is quite authoritarian, for a while it had strong popular support. Indeed, both Brasília and industrial São Paulo were initially built on the basis of massive popular engagement with the project of modernization and belief in progress. This combination of authoritarianism with genuine popular support has a well-established label in Latin American politics, namely, populism. It dominated Brazilian politics from the 1940s until the military coup interrupted it in 1964. The military dictatorship that ruled Brazil between 1964 and 1985 ended popular engagement by political repression. Nevertheless, development continued to be the regime's main objective. Moreover, the same planning and governmental instruments served well the developmentalist policies of the dictatorship. In fact, it was during this regime that development achieved some of its most emblematic marks. This included not only economic growth rates of up to 12 per cent per year, but also the construction of roads and telecommunication infrastructure and the dissemination of social services.

In other words, intense modernization and urbanization in Brazil took place either without popular participation (military regimes) or with elite-controlled popular participation (populist regimes). Missing from all of these governmental rationalities was the project to turn Brazil's masses into modern political citizens who participate meaningfully in political and electoral decisions. As with the polity, so with the society: social inclusion was not one of the objectives of the modernization project. As the military regime often declared, it was necessary 'to grow first to divide the cake later'. In short, authoritarianism and profound social inequality are the marks of modern Brazil.

The Context of Change

The national-developmentalist project of modernization started to crumble in the early 1980s under the influence of contradictory forces. On the one hand, there was a deep economic crisis and the subsequent adoption of so-called 'neoliberal policies'. Not infrequently, the justification for these policies has been the need to put Brazil in tune with the next wave of modernization, that is, the new global configurations. On the other, there were political transformations, especially pressures for social and political inclusion that the urban social movements articulated and that eventually led to political democracy.

Transition to democratic rule in Brazil was a long process. The so-called 'political opening' started in the mid-1970s; the first state governors were elected in 1982; and the first election for president was in 1989. The main mark of democratization, however, was not electoral politics. Rather, it was the explosion of popular political participation and the massive engagement of citizens in debating the future of the country. In Brazil, this mobilization was known as 'the rebirth of civil society'. Two forms of political organization, both originated in São Paulo, were especially important in the transition process: independent trade unions and urban social movements. The latter were crucial for transforming the perception of urban space and including urban citizenship in the agenda of democratic consolidation.

Starting in the mid-1970s, numerous neighborhood-based social movements appeared in the poor urban peripheries, frequently with the help of the Catholic Church (Caldeira 1984). Movement participants, a majority of them women, were new property owners who realized that political organization was the only way to force city authorities to extend urban infrastructure and services to their neighborhoods. They discovered that being taxpayers legitimated their 'rights to the city', that is, rights to the legal order and to the urbanization available in the center. At the root of this political mobilization was the illegal/irregular status of the properties that most had purchased in good faith: public authorities denied them urban services and infrastructure precisely because they considered their neighborhoods illegal. Thus, a central inspiration for these movements was an urban and collective experience of marginalization and abandonment, in spite of individual efforts of integration through work and consumption.

The urban social movements were crucial in the larger opposition that helped end the military dictatorship. The demands of these movements were summarized in the idea that Brazil had to change by becoming democratic and enforcing the rights of its citizens. Accordingly, demands included direct elections (*Diretas Já!*), amnesty for political prisoners and respect for their human rights, revocation of all 'laws of exception' imposed by the military regime, and the convening of a Constitutional Assembly to write a new democratic constitution. Several of these demands were met in the first years of the democratic transition, including the promulgation of a new Constitution in 1988. It was written on the basis of ample consultation with organized popular movements and includes a full set of citizens' rights, from the right to four months of paid maternity leave to the more traditional list of rights to life, freedom of expression, and justice. The 1988 Constitution is a document that interprets citizenship rights in the broadest terms, incorporating what is sometimes called all 'generations' of rights.

While the country democratized, however, the conditions that sustained developmentalism eroded. The mythology of progress started to collapse in the 1980s, in São Paulo as elsewhere in Brazil. It began with what is called the 'lost decade', the deep economic recession associated with changes that significantly transformed Brazilian society and many others in Latin America and around the world. Although this is not the place to analyze these changes in more detail, it is important to mention the most important ones as they affected the metropolitan region of São Paulo in the 1980s and 1990s. These include a sharp decrease in population growth; a significant decline in immigration and increase in emigration, especially of upper- and middle-class residents; a sharp drop in the GNP and rates of economic growth; a drop in per capita income; a deep reorganization of industrial production associated with large unemployment and instability of employment; a redefinition of the role of government in the production and management of urban space; and a significant increase in violence – both criminal and police – associated in part with the restructuring of urban segregation. As a result of the economic crisis and related changes, the distribution of wealth – already bad – worsened and perspectives of social mobility shrank considerably. In the periphery, important aspects of the urban inclusion achieved by social movements eroded (Caldeira 2000: Chapter 6). Many people could no longer afford a house of their own and the reduced horizons of life-chances seemed to preclude even the dream of 'autoconstructing' one. The number of people living in *favelas* in the city increased from 4 per cent in 1980 to 19 per cent in 1993.

One of the most important consequences of this combination of economic and social crisis was that the state abandoned the model of governmentality based on protectionism, nationalism and direct participation in production, that is, the main elements of the modernization project. The policies adopted to deal with the economic crisis – usually indicated by agencies such as the IMF and labeled 'neo-liberal' – resulted in the opening of the domestic market to imported products and in the withdrawal of the state from various areas in which it had traditionally played a central role. These areas included urban services, infrastructure, telecommunications, steel manufacture and oil production.

Privatization became the order of the day, the dominant value of the new logic of governmentality that replaced the modernization project. Privatization has consequences, both economically and socially. It means selling off most of the state-owned enterprises (including those offering basic services such as telephone and electricity) to private interests and using the revenue generated to pay the foreign debt incurred under the previous economic model. It entails cutting state subsidies to national production. It signifies unmaking prerogatives and social rights created both in the corporatist labor legislation of the 1930s and 1940s and in the 1988 Constitution (Paoli 1999). It also means that the state 'contracts out' to private enterprises and privately funded NGOs social services that it used to provide (from the delivery of milk to schools to prison management). Moreover, the state now hires NGOs with public funds to develop policy that government agencies used to produce. In short, privatization undermines various pillars of the developmentalist-modernist project and its type of state. In effect, it subverts the idea that the state governs the nation, and indeed creates a nation in its image by being a direct producer of its public through state-owned and managed industry, state-directed public works and planning and state-provided welfare.

Privatization also influences the space of the city and its everyday practices in decisive ways. Pressured by funding cuts and new laws to balance budget, municipal governments throughout Brazil limited their range of intervention and level of investment in the urban environment. Simultaneously, they called on private citizens to invest in their own space in exchange for fiscal incentives and more flexible building codes. In the periphery, citizens had always invested in their space, but as a result of minimal state investment. Now, however, private investment was becoming a matter of state policy for the whole city. Nevertheless, probably the most important forms of privatization that affected the urban environment related to the startling increase in violent crime and fear (Caldeira 2000). Violence and the inability of the state to deal with it have led people to rely on private security and fortification and to imagine city life in terms of numerous new practices of segregation.

In summary, Brazilian society experienced contradictory processes during the 1980s and 1990s: on the one hand, political democratization and the emergence of new forms of democratic citizenship; on the other, economic crisis, privatization and violence that undermined the former, limited the state, closed urban spaces and reduced possibilities of growth.[14]

Democratic Planning and the Neoliberal State

The 1988 Constitution introduced significant innovations in many areas, including urban policy. These were due to a large extent to the lobbying of organized social movements and civil organizations. During the National Constitutional Assembly of 1986–88, these grassroots forces gathered more than 12 million signatures in support of Popular Amendments, successfully pressuring the state to relinquish its jural monopoly and securing a strong presence in the new constitution. During the next two years, state and municipal constitutional assemblies occurred throughout Brazil with similar results. During these many constitutional assemblies, the demands of grassroots forces converged with legal assistance services. Members of the former brought their specific interests to lawyers of the latter who rearticulated them in terms of proposals for new law. In the process, the social movements became educated in both making and using law. Thus, a new conception of citizenship grounded in the popular construction of the law and the exercise of new kinds of rights through legislation began to take root.

One of the most significant sources of this process of innovation is popular participation in urban reform and municipal administration. Growing out of the National Movement of Struggle for Urban Reform founded in 1986 to influence the federal constitution, this participation has rallied around the principle of 'rights to the city' and around the concept of urban self-management (*auto-gestão*).[15] In major cities, including São Paulo, Porto Alegre, Curitiba, and Recife, it has succeeded in developing this conception of urban citizenship into innovative municipal codes, charters and master plans.[16] We will now analyze two of the most important regulations these efforts produced.

One of the Popular Amendments presented to the Assembly generated the Constitution's section on Urban Policy. Article 182 defines the objective of urban policies as 'to organize the full development of the social functions of the city' and establishes that urban property has a social function. Consequently, it determines that local governments can promote the use of urban land through expropriation, forced subdivisions and progressive taxation so that it fulfills its social function. Article 183 creates *usucapião urbano* (akin to adverse possession) as a means of resolving the predicament of residential illegality that affects so many of the working poor. It establishes the possibility of creating an uncontestable title of ownership for residents who have lived continuously for five years and without legitimate opposition on small lots of urban land. These two articles became the basis for a series of legislated acts, regulations and plans that have since transformed the character of urban policy in Brazil.

The constitutional articles required enabling legislation both to define in more precise terms the concept of 'social function' and to create mechanisms for its implementation. For more than a decade, the National Congress debated this legislation under pressure from the lobby of the National Forum for Urban Reform.

The result is the remarkable *Estatuto da Cidade* (City Statute), federal law 10 257 of 10 July 2001. This legislation incorporates the language and concepts developed by the urban social movements and various local administrations since the 1970s. It is quite unusual in the history of Brazilian urban legislation for at least four reasons. First, it defines the social function of the city and of urban property in terms of a set of general guidelines that are substantive in nature. Second, on that basis, it frames its directives from the point of view of the poor, the majority of Brazil's city dwellers, and creates mechanisms to revert some of the most evident patterns of irregularity, inequality and degradation in the production of urban space. Third, the Statute requires that local urban policies be conceived and implemented with popular participation. Thus, it takes into consideration the active collaboration and involvement of the private organizations and interests of civil society. Fourth, the Statute is not framed as a total plan but instead introduces a series of innovative legal instruments that allow local administrations to enforce the 'social function'. Unmistakably, the City Statute is the result of the insurgent citizenship movements of the previous decades. It is an important indication of one of the ways in which democratization has taken root in Brazilian society and of how the grassroots experience of local administration, legal invention and popular mobilization has made its space in federal law.

Echoing the constitution, the City Statute establishes that the objective of urban policy is 'to realize the social functions of the city and urban property' (Art. 2). Urban policy must do so by following a set of comprehensive guidelines. Among the most important, urban policy must 'guarantee the right to sustainable cities, understood as the right to urban land, housing, sanitation, infrastructure, transportation and public services, work, and leisure for present and future generations' (Art. 2, par. I); use 'planning ... to avoid and correct the distortions of urban growth and its negative effects on the environment' (Art. 2, par. IV); produce a 'just distribution of the benefits and costs of the urbanization process' (Art. 2, par. IX); allow the public administration to recuperate its investments that may have resulted in real estate gain (Art. 2, par. XI); and regularize properties and urbanize areas occupied by the low-income population (Art. 2, par. XIV). By such means, the City Statute clearly establishes the production of social equality in urban space as a fundamental objective of urban planning and policy and, reciprocally, turns planning into a basic instrument for equalizing social disparities and securing social equality.

The Statute also creates powerful instruments to enforce its directives. They are of two types: first, there are instruments of management; second, there are instruments to regulate the use of urban land. The two basic types of innovations regarding management are quite substantial: those requiring popular participation in the formulation and implementation of policies, and those considering that urbanization is to be obtained by cooperation between government and private organizations. Chapter IV of the Statute is entitled *On the Democratic Management of the City*, and its Article 45 presents the boldest formulation of the principle of popular participation:

The management organizations of metropolitan regions and urban agglomerations will include mandatory and significant participation of the population and of associations representing the various segments of the community in order to guarantee the direct control of their activities and the full exercise of citizenship.

Chapter IV establishes that cities must implement a variety of mechanisms to insure this public participation in management, from debates, public audiences, and conferences to popular amendments of plans and laws to a process of participatory budget making.[17] In these formulations, it is evident that the Statute imagines a society of citizens who are active, organized, and well-informed about their interests and their government's actions.

This conception of Brazilian society could not be more different from the one that inspired the modernist-developmentalist master plans. Those plans assumed a backward society of silent and mostly ignorant citizens who needed to be brought into modernity by an illuminated and elite avant-garde.[18] Some of the modernist plans, especially Brasília's, did have social equalization as an objective. But even so, it was one to be imposed, already scripted. It would result from the plans, the values embodied in them, and the built environment they produced. Social equality would not, in other words, result from an exercise of citizenship that would generate the plans themselves. Moreover, the language of the modernist plans was one of development, not citizen rights, and its principal target was underdevelopment, not social inequality. The new model of planning turns this logic of development on its head. In this new formulation, the social is not imagined as something for the plan to produce but is rather something that already exists in organized fashion. This organization will be the basis for the creation of urban space, which will in turn confirm a more equitable and just society. The society imagined by the new model is modern, democratic and plural, although still profoundly unequal. The new plans consider that citizens lack resources, are poor, and have their rights disrespected, but not that they are ignorant, illiterate, backward, incompetent, incapable of making good decision, and so forth. While the old plans supposed that society's needs were modernization, progress and development, the new ones imagine that their needs are citizenship and equality (or at least the abatement of the worst effects of social inequality). They suppose that the majority of the population they address needs their rights, not hygiene. Furthermore, whereas the modernist plans dispensed with any consideration of conflict in the imposition of solutions, the Statute and the other legislation it generated see citizens' interests as different and often contradictory. Therefore, they create mechanisms of conciliation and mediation.

In addition to enacting the principle of direct participation of citizens in managing cities, the City Statute also establishes that the government is no longer solely responsible for the process of urbanization. It thus fractures another fundament of the developmentalist model. The latter supposed that the state is the main (if not sole) producer of urban space – of the legal kind, admittedly, but also of the illegal kind, inexorably. According to the Statute, however, the process of urbanization should entail a balanced cooperation, or partnership, between public and private interests. This reconceptualization of roles is not a matter of democratic change alone. In fact, it is associated probably more with the neoliberal turn of the state, which presupposes a substantial shrinkage in the scope of its interventions, and with the exhaustion of resources to fund investments in urban infrastructure. During the developmentalist years, these resources came especially from international development banks and created an almost unmanageable foreign debt in countries such as Brazil. Today, a good deal of this funding is gone. As a result, under democracy, the Brazilian city has a huge social debt of needs with few resources to address them.

Consequently, administrators search for alternative funding, especially from private sector investments. In addition, they develop new legislative instruments that might simultaneously tax the use of urban space and produce social justice. For example, the City Statute introduces a series of mechanisms to tax real estate profit, force the use of underutilized urban properties, and regularize land occupied by low-income residents. The Statute also incorporates an innovative conception of property rights. It separates the right of property from the constructive potential of urban land, creating the possibility of transferring an owner's right to build. This separation allows the government to sell rights of construction beyond the coefficient (an area limit) permitted in city codes as a means to generate revenue for urbanization projects. There are a host of other innovations, including something called Urban Operations that allow a partnership of public and private interests to 'bend the rules' in delimited areas of the city to achieve certain urbanization purposes, as well as provisions for both individual and collective *usucapião* (the latter in the case of *favelas*) to regularize land ownership among the poor.

The City Statute equips urban government with powerful tools to regulate the production of urban space. However, it conceives these measures quite differently from those of developmentalist plans. The differences are impressive. They concern the general principles that inspire the instruments (social justice and citizenship rights), the conception of how local projects will be created (through the democratic participation of organized citizens and their vigilance over governmental actions), the imagination of how projects will be implemented (the partnership between public and private initiatives), and the restricted nature of the interventions (limited urban operations, actions in 'priority areas' rather than total plans). The City Statute is an instrument of democratic governance. It is based on a democratic conception of Brazilian society, as well as a democratic project for it.

It is hard, however, to predict how it will be engaged by local governments and citizens to change their cities. The legislation is still too recent for us to address the problems of its implementation and its potential to transform the patterns of inequality in Brazilian cities. Nevertheless, it is important to look for indicators of this engagement. For this, we take the case of the city of São Paulo. We will not analyze here its Master Strategic Plan (*Plano Diretor Estratégico*), which is the local application of the City Statute and was signed into municipal law 13 430 on 13 September 2002. The analysis of this plan necessitates its own study given its many innovations, such an environmentalist approach to the city's problems and the intense process of political opposition and bargaining it generated. This process forced many changes in the version proposed by the PT (*Partido dos Trabalhadores*, Workers' Party) administration of the city.[19] Moreover, it is still difficult to anticipate its effects in terms of the production of social justice in urban space. Instead, we consider the use of some of the instruments adopted by the Statute even before they were approved by Congress, implementations that already reveal paradoxical results.

Some Paradoxical Uses of the Statute

Although São Paulo has been a crucial site for the organization of political and social movements that helped to democratize Brazilian society, the city has been largely in the hands of administrations at odds with this orientation. The first democratically elected mayor, Jânio Quadros, who took office in 1986, was an old-time conservative populist. During the next term (1989–92), the city was administered by a mayor from the PT, the political party connected in the most direct way with the interests of the working classes and its social movements. However, after this administration of Luiza Erundina, the city had two mayors from the PPB (*Partido Progressista Brasileiro*), a center-to-right and conservative party associated with the real estate and construction industries.[20] In 2001, another mayor from the PT (Marta Suplicy) took office. All these administrations used at least some of the instruments incorporated into the Statute. Following their different uses allows us to discuss some of the paradoxical ways in which democratization and neoliberalization have intertwined in the production of urban space.

During the whole democratic period, one mayor after another developed master plans for the city of São Paulo that never passed City Council because they never generated enough support.[21] These plans were intended to substitute the modernist-developmentist master plan and zoning code passed in 1971, the PDDI (*Plano Diretor de Desenvolvimento Integrado*). The only plan to pass City Council, in 1988, was approved by default, and its legitimacy has always been questioned. As the government retreated under the mantle of neoliberal policies, and as the City Council failed to approve one master plan after another, contingency planning ruled the city. This meant contradictory initiatives. One the one hand, various administrations were able either to introduce or to use a few instruments that are similar to those of the City Statute. On the other, organized private interests moved in to fill the space opened by the withdrawal of the state. The city of São Paulo of the last 15 years is a city where private investors intervened decisively, sometimes in partnership with local government, to improve the areas of their investment with the objective of increasing significantly the value of their real estate. One of the results of this action is the consolidation of a new pattern of urban segregation based on the proliferation of fortified enclaves, that is, of privatized, enclosed and monitored spaces for residence, consumption, leisure and work.[22]

Policies to tax real estate profits and to attract private investment in urbanization are not inventions of the City Statute. Rather, they have been practiced for some time in São Paulo and other cities. We look at two such instruments used in the last 15 years in São Paulo: the so-called 'paid authorization' (*outorga onerosa*) and the Urban Operations. Paid authorization refers to the possibility that the government may sell rights of construction beyond the coefficient allowed in city codes, if it uses the funds thus generated for urbanization projects. Urban Operations are projects to preserve, revitalize and/or transform specific urban areas, through partnerships of public and private investment. These operations must be defined by law, and the norms that regulate them may differ from those of the rest of the city. Paid authorization is a core instrument of Urban Operations. Both have been introduced in the administration of São Paulo as means of revising the role of the state in the production of urban space and of fulfilling the need to find new forms of investment in urbanization. In some cases, the objective was to produce social justice and allow the administration to recuperate investments that produced real estate gains; in others, it was to benefit real estate investors. The results of the latter deepened spatial segregation.

The idea of paid authorization was first introduced in São Paulo in 1976 during the administration of Olavo Setúbal (Câmara dos Deputados *et al.* 2001: 68). Although it was not transformed into legislation at that point, it was incorporated into the discussions of the social movements and organizations addressing the urban question since the start of the democratization period.[23] What appealed to these democratic interests was the possibility of generating new sources of revenue for urban development. However, when the idea was first transformed into law during the administration of Jânio Quadros, with the name *Operações Interligadas*, it had an unexpected twist.[24] It allowed the government to offer private developers the right to build beyond the limits set in zoning codes in exchange for their private investment in 'popular housing'. Such operations were conceived in the context of Quadros' plans of *desfavelamento*, that is, the removal of *favelas* and their population, especially in central areas. Proprietors of areas occupied by *favelas* could petition the city to change the rules of use and occupation in any land they owned in exchange for the construction of popular housing. This conception was at odds with the most common interpretations of paid authorization, according to which the instrument should apply only to specific areas of the city selected on the basis of urbanistic projects, such as those to increase urban density in areas of good infrastructure. In Quadros' interpretation, however, the bending of zoning rules was particularistic, for it did not follow any specific urban project, but rather applied to any area in the city where a *favela* might exit. In fact, it was an instrument of social segregation. The City Statute later discarded this particularistic use of paid authorization. Instead, it adopted the conception developed by the PT administration of Luiza Erundina under the label *solo criado* (created soil), which required the use of urban projects to designate areas of the city eligible for paid authorization.[25] Paulo Maluf's and Celso Pitta's PPB administrations subsequently used this instrument in conjunction with Urban Operations.

In São Paulo, Urban Operations were introduced in the mid-1980s and used by the conservative administrations that followed.[26] In general, operations launched in the 1990s either failed to transform their areas or generated further social inequality, segregation and real estate profit. Three operations – Anhangabaú, Centro and Água Branca – were located in deteriorated downtown areas. Each resulted basically in only one private project. The third seems the most successful but has been limited to creating the infrastructure needed by the only private project approved for the area.

Two other Operations – Faria Lima and Águas Espraiadas – are in the area of the newest business districts along the Pinheiros River. They were designed to install the kind of infrastructure required for the development of 'intelligent' office complexes and accompanying commercial malls and residential units (closed condominiums) for their workers. The Faria Lima Operation is a clear example of the risks of one of the determinations of the Statute: that the funds raised by an urban operation should be used exclusively within areas of its jurisdiction. Since Faria Lima is a region of high real estate values, further investment has only augmented its privileges. Moreover, because the Operation encouraged the aggregation of lots, it had a strongly regressive impact in the real estate market, expelling modest investors and discouraging small-scale use. Thus, the Operation transformed a residential area of small lots into a business area of large developments. Similar effects happened in the adjacent area of Berrini/Águas Espraiadas, which received large investments in road construction and river channeling. Moreover, this area benefited from an infamous partnership between the city under the administration of Paulo Maluf and private investors. The agreement put together city agencies of social work and a pool of enterprises, the offices of which were located in the operation area of Berrini Avenue. The objective of the partnership was to remove a *favela* near the offices. Berrini Avenue became one of the most fashionable addresses for business in the city during the 1990s, and its poor neighbors were viewed as an eyesore. Although many *favelas* had been displaced in the city before, this was the first time in which representatives of the private sector participated directly in the removal. Although they used a philanthropic discourse to legitimate their initiative, they never disguised their obvious objective of obtaining real state valorization. Similarly, the city did not disguise its interest in the partnership.[27]

In summary, the Urban Operations combining public and private investors in São Paulo have thus far increased inequality and spatial segregation. The urban areas that they re-qualified are emblematic of new trends in segregation transforming the city in the last two decades.[28] Clearly, once social agents engage them, instruments of planning and governmental regulation do not necessarily produce the results their formulators intended. Brasília is a clear example in this regard, as Holston (1989) demonstrated. For the United States, Mike Davis (1990) gives a compelling analysis of how NIMBY movements in Los Angeles have used democratic instruments to produce exclusion and segregation. These examples only make us skeptical about what to expect from some of the new instruments of urban management. They also force us to consider the complex relationship between democratic and neoliberal planning.

Undoubtedly, in the last 20 years, city administrations in Brazil have reconceptualized the role of the state, the nature of planning, and the relationship between public and private sectors in the production of urban space. The results have significantly transformed the dominant modernist-developmentalist model of planning and urban management. Undoubtedly, too, democratization alone cannot explain these innovations. Indeed, the interconnections between democratic and neoliberal rationalities of government are intricate, yet still underinvestigated. Although many new instruments have been introduced in the name of an expanded role for 'civil society', this role has in fact often only guaranteed specific private interests, as in São Paulo's Urban Operations, instead of a broad representation of different perspectives. To date, however, these operations have been implemented by administrations that disregarded the practices of participatory democracy and interpreted the partnership of public and private in predominantly neoliberal terms, as a means to realize market interests and not social justice. Nevertheless, as we have shown, the new planning initiatives have the potential to generate urban spaces that are less segregated and that fulfill their 'social function' – spaces that are, in short, more democratic, in the sense that their resources are equitably distributed and their citizens are active participants in their making and management.

Therefore, one can hope that an administration committed to those ends will succeed in using the new instruments of planning to realize them. This expectation has some legitimacy, for the democratic practices of popular social movements and local administrations have already transformed the modernist model of urban planning and government into the vastly more democratic project embodied in the City Statute. That is no small achievement.

Acknowledgements

The authors are grateful for research support from the Núcleo de Estudos da Violência (Universidade de São Paulo and FAPESP), a J. William Fulbright Foreign Scholarship, and a Fulbright-Hays Faculty Research Fellowship. A previous version of this chapter appeared as: Teresa Caldeira and James Holston. 2005. 'State and Urban Space in Brazil: From Modernist Planning to Democratic Interventions'. In Aihwa Ong and Stephen J. Collier, (eds), *Global Assemblages: Technology, Politics, and Ethics as Anthropological Problems*, London: Blackwell, 393–416.

Notes

1   *Favela* refers to a set of shacks built on seized land. Although people own their shacks and may transport them, they do not own the land since it was illegally occupied. From the point of view of urban infrastructure, *favelas* are extremely precarious. The shacks are close together, there is no sewage service and frequently no piped water, and generally people obtain electricity by illegally tapping into existing electric lines. *Cortiço* is either an old house whose rooms have been rented to different families, or a series of rooms, usually in a row, constructed to be rented individually. In each room a whole family sleeps, cooks and entertains. Residents of various rooms share external or corridor bathrooms and water sources.

2   For further analyses of Brasília, see Holston 1989 and 2001.

3   Brazil's urban population represented 36 per cent of the total population in 1950, 68 per cent in 1980 and 81 per cent in 2000 (in a total population of almost 170 million). In 1980, Brazil already had nine metropolitan regions with populations over one million.

4   For an analysis of modernism and modernization in Brazil, as well as of the creation of Brasília and Kubitscheck's Plan see Holston (1989). For an analysis of the industrialization of São Paulo see Dean (1969) and Singer (1984). For analyses of the transformations of this city during the developmentalist period, see Morse (1970: Part IV) and Meyer (1991).

5   For an analysis of this belief in progress and its social consequences especially for the case of São Paulo, see Caldeira (2001).

6   See Holston 1989: 31–58 for a discussion of the Brazilian embodiment of the CIAM model city.

7   For a discussion of these strategies and of the residents' reaction to them, see Holston (1989: 163–87). The reduction both of family social space and of the expression of individuality in residential architecture is consistent with modernist objectives to reduce the role of private apartments in the lives of residents and, correspondingly, to encourage the use of collective facilities.

8   For a full account of the Brazilianization of Brasília, see Holston 1989.

9   The *Plano Piloto* was planned for a maximum population of 500 000. As of 2000, the date of the most recent findings, it has a population of 198 400. If we include the Lake

districts North and South, we add another 57 600 residents, for a total that is still just half Brasília's planned population. Moreover, the demographic imbalance between center and periphery has only worsened with time. At inauguration, the *Plano Piloto* (including the lake districts) had 48 per cent of the total Federal District population and the periphery (both Satellite Cities and rural settlements) had 52 per cent. In 1970, the distribution was 29 to 71 per cent; in 1980, 25–75 per cent; in 1990, 16–84 per cent; and in 2000, 13–87 per cent. Source: IBGE-CODEPLAN 2000.

10   The urban plans for São Paulo of the 1950s to the 1970s were modernist and developmentalist. These types of plans continued to be produced well into the 1970s. The clearest example is the integrated plan of development approved in 1971 (*Plano Diretor de Desenvolvimento Integrado*).

11   The mechanisms that created a legal/illegal city started to appear in São Paulo at the beginning of the 20th century and were constitutive of Brazilian land occupation and legislation since early colonial times (Holston 1991b). In the case of São Paulo, legislation during the 1910s established a division of the city into four zones: central, urban, suburban and rural. Most of the planning statutes created at that time applied only to the central and urban zones, leaving the other areas (where the poor were already starting to move) unregulated. When some legislation was extended to these areas, such as requirements for registering subdivisions and rules for opening streets, it did not take long for developers to gain exemptions. The requirements that new streets had to have infrastructure and minimum dimensions, for example, could be legally bypassed after 1923 when a new law offered the possibility of creating 'private streets' in suburban and rural areas. The legal rules for the urban perimeter did not apply to these private streets. Probably the best example of this mechanism of exception relates to the required installation of infrastructure that, starting at the beginning of the century, depended on the legal status of a street. Most of the new streets, especially in the suburban and rural areas, were either irregular or illegal, and therefore exempted from this requirement by definition. Given the intense settlement of urban migrants in these areas, this exclusion amounted to an extraordinary subvention for developers and hardship for new residents. Although the new subdivisions were progressively legalized and given urban status through various amnesties (1936, 1950, 1962, 1968), these decrees were each ambiguous enough to leave to executive discretion the determination of which streets fit the criteria for legalization, and therefore for urban improvement, and which did not. For detailed analysis of this mechanism and its effects on São Paulo's legislation and urban space, see Holston 1991b, Caldeira 2000: Chapter 6, and Rolnik 1997.

12   The population of the city grew from 579,033 in 1920 to 3,781,446 in 1960, according to the census. In 2000, it was 10 405 867 in the city and around 18 million in the metropolitan region (the combined area formed by the city plus 38 surrounding municipalities). Population density in the city dropped from 11,000 inhabitants per sqkm in 1914 to 5,300 in 1963. In 2000 it was 6,823 inhabitants/sqkm.

13   Until 1985, illiterate people in Brazil (all from the working classes) could not vote. Moreover, the military regime that took power in 1964 eliminated all elections for executive offices.

14   The contradictions between an 'insurgent democratic citizenship' and a 'disjunctive democracy' in Brazil are the focus of a forthcoming book by Holston. It also analyzes what we discuss in the next sections of this chapter, namely, the emergence of new forms of citizenship in relation to law and its institutions, and the transformation of the insurgent notion of 'rights to the city' developed by the urban social movements into new modes of urban planning.

15   This movement was later consolidated into the National Forum of Urban Reform, which congregates numerous NGOs, social movements, and trade union organizations interested in urban reform. The Forum is still quite active in promoting urban legislation at all levels of government.

16 A discussion of some of these innovations is found in Silva 1990.
17 Most of these procedures have been used by local administrations, especially from the PT (Worker's Party) for at least 15 years. They became standard for any administration that wants to be recognized as popular. The participatory budget process is a mechanism for the formulation of the annual city budget through public audiences in which neighborhood and district representatives have the right to voice and vote.
18 It is worth remembering that one of the arguments that justified the prohibition of direct elections after the 1964 military coup was that people (meaning poor people) did not know how to choose and to vote and should therefore be governed by those who know.
19 São Paulo's Master Plan has 308 articles dealing with not only urban policies per se but also with issues ranging from the rights of minorities to employment. It is a clear example of how the experience of social movements and of democratic local administrations (mostly from the PT) have framed conceptions of urban management in contemporary Brazil. This Master Plan incorporates the language and the instruments of the City Statute as well as a whole new series of concepts and initiatives developed by the social movements and Forums such as partnerships, solidarity development, project incubators, participatory budget, and so on. The consideration of the Plan by City Hall involved intense debate – as expected – and considerable opposition, especially from organized groups eager to defend their real estate interests. A number of the innovations in the 2002 Master Plan had already been introduced in previous plans that did not pass City Council, such as the Master Plans proposed by mayors Mário Covas in 1985 and Luiza Erundina in 1991.
20 The mayors were Paulo Salim Maluf, who had previously served as non-elected mayor and governor during the military dictatorship, and Celso Pitta.
21 For an analysis of the principal master plans of the city of São Paulo in the 20th century, see Somekh and Campos (2002).
22 For a full analysis of the consolidation of this new pattern of urban segregation and of the context of increasing violent crime and fear in which it occurs, see Caldeira 2000.
23 One account of the transformations of this notion, as engaged by various social movements and forums of urban reform, is given in Câmara dos Deputados *et al.* 2001: 68–71. See also Somekh (1992).
24 Municipal Law 10 209 from 1986. Because changes in zoning were not authorized by City Hall, these operations were later prohibited by the justice system under the allegation that they were at odds with the state constitution.
25 The 1991 Master Plan elaborated under Erundina's administration, but not approved, was the first to follow the new Constitution's principles on urban policy. The Plan reaffirmed the social function of the city and of urban property and proposed to substitute the existing zoning code with the *solo criado* rule. It recommended that the whole city have the same utilization rate (*coeficiente de aproveitamento*) of one, instead of the multiple rates defined by PDDI. The utilization rate defines the relationship between the total permitted built area and the total area of the lot. The right to build above the rate of one-to-one would have to be purchased from the city. Furthermore, the Plan determined the areas and the quantities permitted for such purchases. This same principle of a single utilization rate was reintroduced in the proposal for the 2002 Master Plan. However, to get it passed by City Council, the administration had to negotiate the single rate and raise its limits.
26 The Master Plan 1985–2000 elaborated by the Mário Covas administration defined the possibility of Urban Operations.
27 The transformations of Berrini Avenue and the areas around Águas Espraiadas and Faria Lima, as well as the removal of the nearby *favelas*, are analyzed by Fix (2001) and Frúgoli (2000).
28 See Caldeira (2000: Chapters 6–8) for a fuller discussion of this kind of spatial segregation.

Chapter 9

The Chronic Poor in Rio de Janeiro: What has Changed in 30 Years?

Janice E. Perlman

Brazil has changed dramatically over the past 30 years. A gradual political *abertura* (opening), starting in the late 1970s, led through a series of incremental steps to the end of the dictatorship in 1984 and the re-democratization of the country. However, the 'economic miracle' of the 1970s also gave way to triple digit inflation during the 1980s, then to stagnation and a series of devaluations of the currency. Efforts to curb inflation culminated in the *Plano Real* (Real Plan) established by Fernando Henrique Cardoso in 1994, but this did not solve the problem of economic growth, which remained low during the 1990s, resulting in increasing unemployment and inequality over the decade. For the urban poor, marginality has been transformed from a myth to a reality as high hopes for a better life for their children have been dashed by a series of barriers. Brazil continues to be one of the most economically unequal countries in the world with the top 10 per cent of the population earning 50 per cent of the national income, while about 34 per cent of the population lives below the poverty line.

This chapter[1] is based on longitudinal survey data, life histories and participant observation in three low-income communities in Rio de Janeiro, initially collected by the author in 1968–69, and again 30 years later in 1999–2003. The trajectories of families and individuals were followed from fishing and agricultural villages to *favelas* (squatter settlements) and *loteamentos irregulares* (illegal or cladestine low-price subdivisions lacking urban services and infrastructure), in the three areas of Rio de Janeiro where poor people could then live. These areas studied were: Catacumba, a *favela* in the wealthy South Zone that has since been removed and its residents relocated to more distant public housing; Nova Brasília, a *favela* in the industrial North Zone that is now a battleground between police and drug traffickers; and Duque de Caxias, a peripheral municipality in the Fluminense Lowlands (Baixada Fluminense), where three *favelas* and five *loteamentos* were selected for study. These communities as well as the housing projects to which the Catacumba residents were removed – Guaporé-Quitungo and Cidade de Deus (City of God), which is now famous for the movie adaptation of the book authored by Paulo Lins – are shown in Figure 9.1.

Despite three decades of public policy efforts in Brazil, first to eradicate *favelas*, then to upgrade and integrate them into the city, both their number and the number of people living in them has continued to grow. There were approximately three hundred *favelas* in Rio de Janeiro in 1969 and today at least twice as many. Not only have *favelas* increased in number and size, they have merged to form vast

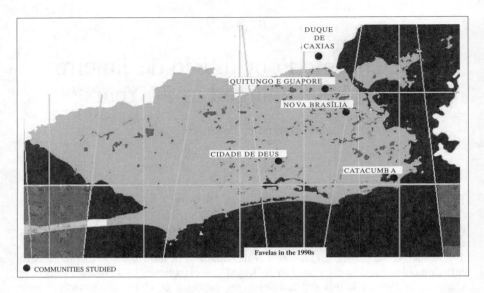

Figure 9.1   Map of the communities studied
Source: Pró, URB.

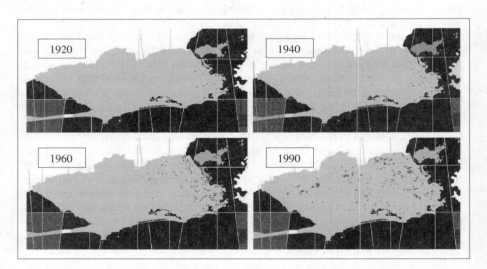

Figure 9.2   Evolution of *favelas* of growth in Rio de Janeiro (1920–1990)

contiguous agglomerations or 'complexes' of communities across adjacent hillsides as seen in a series of maps from 1920, 1940, 1960 and 1990 (see Figure 9.2).

Statistics shown in Table 9.1 indicate that in every decade between 1950 and 2000, Rio's *favela* population grew much more rapidly than the city as a whole with the exception of the 1970s when *favela* eradication programs forcibly removed over 100,000 people into public housing or sent them back to the countryside. Most striking is that during the period 1980 to 1990, when the overall city growth rate dropped to 8 per cent but *favela* populations surged by 41 per cent. Then, from 1990 to 2000, when the city's growth rate leveled off at just less than 7 per cent, *favela* populations continued to grow by 24 per cent resulting in a current all-time high percentage of Rio's population.

Table 9.1   Growth rates of *favelas* and Rio de Janeiro city population by decade (1950–2000)

Year	Favela population	City population Rio de Janeiro	Favela pop./ City pop.	Growth rate Favelas	Growth rate Rio population
1950	169,305	2,337,451	7.24%	–	–
1960	337,412	3,307,163	10.20%	99.3%	41.5%
1970	563,970	4,251,918	13.26%	67.1%	28.6%
1980	628,170	5,093,232	12.33%	11.4%	19.8%
1990	882,483	5,480,778	16.10%	40.5%	7.6%
2000	1,092,958	5,857,879	18.66%	23.9%	6.9%

It is also clear that *favela* growth has not been spread evenly over the metropolitan region, growing 108 per cent in the West Zone over the past decade in response to greater availability of vacant land, newly accessible transportation and the burgeoning upper class developments in the Barra de Tijuca, as compared with 21 per cent and 14 per cent respectively in the already consolidated South and North Zones. This growth is evident in Table 9.2, both in terms of number of *favelas* and population in the *favelas*.

Table 9.2   Growth rate by zone from 1980–1992

Zones	Number of Favelas			Favela population		
	1980	1992	Rate of growth	1980	1992	Rate of growth
South	25	26	4%	65,596	79,651	21%
North	22	25	14%	49,042	55,768	14%
West	86	195	127%	94,002	195,546	108%
Suburbs	194	270	39%	416,307	532,340	28%
Central	45	57	27%	92,119	99,488	8%
Total	372	573	54%	717,066	962,793	34%

The fact that *favela* growth has been outpacing city growth was initially due to the influx of migrants from the countryside, particularly the Northeast, Minas Gerais, and the State of Rio de Janeiro itself. But today, the rates of rural-urban migration are much lower, and the West zone is the main reception area for significant numbers of newcomers to the city, indicating that the growth is due to a combination of natural increase[2] and some immigration from other areas within Rio de Janeiro and from other states.

Methodology

In the original study from 1968–69, we selected 200 men and women 16–65 years of age at random from each of the three communities (i.e., the random sample, 600 in total) and 50 leaders selected by position and reputation in each (i.e., the leadership sample, 150 in total). When we went back in 1999, we were able to relocate 41 per cent of the 750 people in the original study. Because we did not record the names of the interviewees of the random sample in order to protect their anonymity but had the last names of the leaders, which were also better known, we were able to locate the latter much more easily – finding 61 per cent of the leadership sample vs. 36 per cent of the random sample.

As shown in Figure 9.3, Catacumba, where we expected the lowest success rate in finding our original interviewees – since forced eviction in 1970 had scattered families across several distant housing projects – turned out to have the highest rate, due to the strong sense of solidarity created through years of struggle for collective urban services, culminating in the long battle against eviction. By contrast, the lowest relocation success rate was for original interviewees from Duque de Caxias – particularly in the *loteamentos* (rented or privately owned lots). Contrary to predictions in the literature that home ownership would create stable settlement patterns, we found a much higher turnover rate among those living in *loteamentos* in Caxias than in *favelas*. This could be explained by a variety of factors including the fact that many of the people we interviewed in the *loteamentos* were renters and not owners, that more of them moved out, and that since there were weaker social ties and fewer community organizations, the neighbors and local establishments had no ongoing contact with those who had left.

Although it is impressive to find 41 per cent of the study participants so many decades after the original research, it raises the question of what happened to the other 59 per cent and whether our findings can be generalized. It is conceivable that we found only the poorest as the better off had moved away, or that we found only the most successful as the worst off had been forced to live on the streets or under bridges and were, therefore, not located. Thus, considerable effort was spent testing for bias. We did this by using the 1968–69 data to compare those whom we found, either alive or deceased and those we could not find.

We found no systematic bias by any measure of socioeconomic status, including income, educational level, possession of household goods, people per room, or an index we created combining these measures among those we were able to locate and those we could not. Nor was there any difference according to race, manual vs. non-manual occupation, or the level of urban services in the household. Comparing those

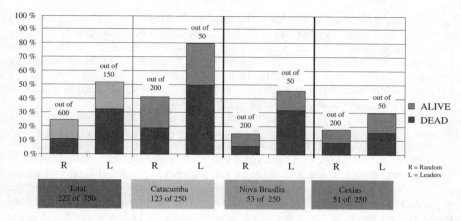

Figure 9.3   Percentage chart of our interviews by community, by random or leadership sample, and by 'survival'

we were able to find with those we could not locate, there was also no difference in gender or age, but women and the youngest cohort were overrepresented in those found alive as would be expected.

The main factors determining our chances of success in relocating the people were (1) of the degree to which they were members of some community group or more connected socially to networks of friends and relatives, and (2) that they lived in a *favela* and not a *loteamento* in Caxias. As explained above, this is because people in the *loteamentos* did not have to struggle collectively to get community services and to fight against removal as those in *favelas* did, and therefore their ties to the community were weaker, and when they moved, there were no family members, friends or community leaders who kept track of them as there were in the *favelas*. For the people who had died in the interval between the two studies, mostly the eldest of the random and elite samples and more men than women, we located their spouse or eldest child and asked them to complete the life history data for the original interviewee.

Analysis of Findings

Where are They Today?

In a model of chronic poverty one might expect that the majority of people who were living in *favelas* in 1968–69 would be living in *favelas* in 2003. This was true for only one third of the random sample (33 per cent; see Table 9.3). Another 27 per cent are still in the public housing projects to which they were removed in 1970, and it appears that although this was highly detrimental to their lives at that time, it has improved their lives overall. The most encouraging finding, however, is that 40 per cent of those who were in *favelas* at the time of the original study are now renters or owners of houses or apartments in legitimate neighborhoods (*bairro*). Only a handful have made it to the South Zone, but many have either bought land in peripheral parts of the city and constructed on it or managed to rent apartments in the less expensive parts of the city. In the random sample of their children, the pattern was similar: 35 per cent are now living in a *favela*, 21 per cent in a housing project and 44 per cent in a legitimate neighborhood.

The leaders have done even better: only 13 per cent of them have remained in *favelas*, 32 per cent are in public housing, and over half or 55 per cent are in legitimate neighborhoods. In many cases, these leaders have bought fairly large lots of land in distant areas (i.e., the outskirts of Niterói, the less-settled areas of Jacarepaguá, Bangu, Campo Grande, or the peripheral neighborhoods of Caxias) and have built family compounds. These houses are generally not visible from the unpaved road, and it is hard to find them. Once you go through the metal gate into the driveway, however, there is often a little garden area and veranda inside and a series of attached and separate houses, the main one usually two or three stories high. They are self-built over time (*autoconstrução*) as the original *favelas* were, and as the family acquires capital, it is often reflected in the tiling of the kitchen and bathroom or the furnishing of the bedroom. Many families said that they had purchased their major home improvements during the period just after the 1994 *Plano Real*, when the local currency was pegged to the value of the dollar and its purchasing power suddenly much greater.

This move from *favela* to *bairro* (legitimate neighborhood) is a major status leap as well as a geographic change: the distance from the center and the incurring of the costs of monthly rent or by purchasing land and building materials is traded off against the stigma and danger of living in a *favela*. Some of the families I interviewed who had left the *favelas* or the public housing projects to escape from the violence and distance their children from the drug trafficking scene, reported feeling very lonely and isolated in the apartments they had rented and financially squeezed by the monthly rent and utility charges.

Table 9.3   Current location of *favela* residents interviewed in 1969

Type of sample	Type of community		Legitimate neighbourhood	
	Favela	Public housing project		
Random (alive)	33%	27%	40%	100% (N=115)
Leaders (alive)	13%	32%	55%	100% (N=38)
Children	35%	21%	55%	100% (N=119)

Livelihood and Well-Being

How are the people originally interviewed living today? Due to the advanced age of the group, a majority (60 per cent) are living on federal retirement payments. They receive about one 'minimum salary' per month (equivalent to about US$90). In many cases this is the main source of income for the entire household, often consisting of several unemployed children and young grandchildren. The pension is sometimes supplemented by charity and scrounging such as the *cesta básica* (basic-needs basket) distributed by religious groups or gathering leftover produce from local market stands. More importantly, under the government of President Fernando Henrique Cardoso, a number of innovative poverty programs were established, which give small cash transfers to poor families in exchange for each child from 7 to 14 who is kept in school, is inoculated and can prove regular medical check-ups. An array of such programs (*bolsa escola*[3] and others) have been consolidated as the Family Grant Program and expanded by the Workers' Party (PT) President, Luiz Inácio 'Lula' da Silva. This is done through a type of debit card for eligible families and is now providing an average monthly benefit of about US$24 (see Dugger 2004). Such government programs play an important role in the survival of these families, which is why theories of 'advanced marginality',[4] which posit a retrenchment of the welfare state do not hold true for the case of Brazil. In fact, in Brazil the welfare state is now expanding.[5]

Socioeconomic Status

In order to look at the relative well-being of the individuals in our study, we created an index of socioeconomic status (SES) using income, education, household goods and people per room. This enabled us to create a score for each person and to cluster these scores as 'high', 'medium' and 'low'. From the literature on chronic poverty, the feminization of poverty and race-based poverty, we expected that race and gender and particularly the gender of the household head would be strong determinants of the level of SES.

In the 2001 data, neither race, nor gender, nor gender of household head were correlated with SES. It is of interest, however, that in 1969, race was significantly correlated with SES. Forty-two per cent of whites, 34 per cent of *pardos* (*mulattos*) and only 6 per cent of blacks were in the high SES category at that time. Today, however, there is no correlation, and in fact, the racial group with the highest percentage in the top SES category is not whites (28 per cent) but *pardos* (41 per cent); blacks are in-between (33 per cent). More work must be done on this, but one possible explanation is that the best and brightest among whites had other opportunities and were not the ones who tended to be living in a *favela* in 1969. This may explain why the whites did more poorly than *pardos* and blacks over time. On the other hand, the pervasive power of racial prejudice may be seen in the percentages of people of each skin color who have managed to move out of *favelas* or public housing projects (or *conjuntos*, i.e., high-rise walk-ups) into legitimate neighborhoods. Forty-three per cent of the whites, 30 per cent of the *pardos*, and only 24 per cent of blacks now live in legitimate neighborhoods.

Looking at gender, there was not a statistically significant difference between males and females in the high SES category (24 per cent of males and 37 per cent of females), although it is of interest to note that there was a higher percentage of females. Likewise, there is no significant difference between the SES of male- and female-headed households; but in this case, slightly more male-headed households

(39 per cent) than female-headed households (31 per cent) ended up in the top SES group.

Of over 80 variables tested, only two were strongly associated with SES: the type of community the person was living in – *favela*, *conjunto* or *bairro* – was strongly associated with SES, as was the fact that a person remained in the original community as opposed to having moved away. Almost twice as many people who are living in *bairros* were in the top SES category (44 per cent) as compared with those in either *favelas* (10 per cent) or *conjuntos* (41 per cent each).

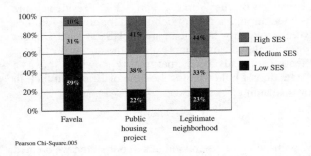

Pearson Chi-Square.005

Figure 9.4   Chart of socio-economic status by type of community

It appears, however, that those in *conjuntos* are better off, since when we look at residential mobility among those who never moved, only 18 per cent are in the high SES group, compared with 38 per cent who went to a *conjunto* and stayed there and 39 per cent of those who moved at all, regardless of where they ended up. As a result, the group one might consider as the chronically poor are those who lived in a *favela* all of their lives and were unable to get out. This is consistent with the discourse on marginality and exclusion.

Intra-generational Mobility

Some of the people we found had improved their levels of living over their lifetimes, some had stayed the same, and some suffered a decline. In order to measure socioeconomic mobility, we compared standardized measures of current and past socioeconomic status[6] using each person's SES scores in 1969 and in 2001. This is only a relative measure and only relates to two points in time. There could have been many ups and downs in the intervening years,[7] but for now, it is a relative comparison between two fixed points in time.

According to these measures, we found that, within our sample, the leaders who already had higher SES than those in the random sample in 1969 also had greater relative upward mobility. We also found that all those who moved out of *favelas* into legitimate neighborhoods showed higher socioeconomic mobility along with physical mobility. This may be interpreted to mean that the ones most able to move out did so thereby hampering their future chances of success, creating a vicious cycle. Those who remained in the *favelas* had the lowest SES, even lower than those who were forced to go into housing projects.

As with SES, the lack of correlation with race, gender, or gender of the household head was striking. Only two variables from among more than 80 tested correlated significantly with mobility. The first correlation reconfirms the findings reported above: there is a strong positive correlation with residential and socioeconomic mobility. Among those who moved their place of residence, 33 per cent had relative upward mobility comparing their SES in 1969 and in 2001, as opposed to those who stayed in the *favelas* where only 18 per cent were in the upward mobility group. Correlation does not imply causality or explanation, but what is probably true is that those who could move did so and once out of a *favela*, had more life opportunities to improve in multiple ways. We will return to this discussion when we look at the issues of barriers to mobility and the stigma of living in a *favela* as opposed to the so-called *asfalto* (asphalt) or legitimate city.[8]

The other significant correlation is with social capital, particularly with membership in community groups. This confirms and complements the literature, which posits the importance of social networks, organizational membership, and density of social relationships for economic prosperity and political stability at the regional level (Putnam 1994) and of those same variables for well-being among the poor. Even before the term 'social capital' had come into vogue, experience had shown that mutual help through networks of reciprocal exchange was a major coping and survival mechanism of the poor (see, e.g., Lomnitz 1977; Campbell 2003; Perlman 2004).

In the case of the *favelados* (residents of a *favela*) we studied, almost 60 per cent of those who participated in one or more community association vs. 17 per cent of non-participants had relative upward mobility. That may be explained by the fact that those who were better off to begin with, and therefore had the time and resources to participate, were in a better position to be upwardly mobile. One could also conclude that those who had the motivation, hope and energy to belong to a community group and advocate for the community's improvement were more likely to succeed in their lives. In any case, the relationship holds up even when controlled for levels of income, education and optimism.

There is also a relationship, although not statistically significant, between relative upward mobility and two other indicators of social capital: friendship and kinship networks, and trusting social relations with neighbors. Among those who report having many friends and relatives in the community with whom they interact frequently, 42 per cent scored high in relative upward mobility as opposed to only 23 per cent among those who are more isolated. In terms of trusting social relations with neighbors, 47 per cent of those who felt they 'could trust most or all of their neighbors' vs. 25 per cent of those who 'could trust few or none' demonstrated relative upward mobility.

The link between relative upward mobility and reciprocal helping networks may be weakened by the lack of family income. As Mercedes de la Rocha (1994) has pointed out, there has been a change in the past decades from 'the resources of poverty to the poverty of resources'. Her work in Guadalajara, Mexico, showed that once chronic poverty becomes endemic in a family and no one in the household has any outside source of revenue, these networks tend to break down due to lack of reciprocity. This remains to be tested for the Rio de Janeiro case.

Inter-generational Mobility

This research project, when completed, will have data on four generations, spanning the end of the 19th century into the 21st century. We will have collected information about the original 750 interviewees, their parents, their children and their grandchildren. As interviews with the grandchildren are still being conducted, the comparisons here will focus mainly on the original interviewees and a random sample of their children. Three aspects of mobility have been selected: consumption, education and occupation.

Consumption

Consumption is increasingly being used as an indicator of well-being as income has become controversial as a measure. Income was seen as easy to obtain, objective and convenient in comparisons, but this has been challenged by social scientists and even economists who have found it does not provide a meaningful indicator of well-being. In part, this is because of differences in purchasing power in different places at different times, the varying values of the currency, the disregard for non-monetary benefits and the absence of assets in the equation. This last point was strikingly brought home to me many years ago when I witnessed a clerk at Sears denying a credit card application from a well-known person who was independently wealthy. The problem was that on his application, he had filled out a sum in the monthly expenditure column far exceeding the sum (zero) he had entered in the column on monthly income from his present job. The question of assets was not included in the application form. Recently Moser *et al.* (2001) have argued for the inclusion of assets and livelihoods in economic measures at the other end of the spectrum, that is, poverty assessment. For the urban poor, who rarely have any savings or investments, their 'net worth' may be seen, in part, by the condition and materials of their houses, their access to urban services and their possession of household goods. Our findings from the Rio de Janeiro re-study show an astonishing increase in consumption on all of these measures in the past 30 years.

Taking infrastructure and housing conditions as an example in 1969, only 27 per cent of *favelados* had running water in their houses, 52 per cent had bathrooms inside their houses (although less than half were connected to any sewage system), and none had a legal electricity connection, although 66 per cent had access to some illicit form of electricity. Today, access to these amenities is almost universal. This change has revolutionized the quality of life, especially for the women who, when I lived there, had to wake well before dawn to line up at the collective standpipe to fill their square 5 gallon cans with water (sometimes only flowing at a trickle) and carry them back up the hill on their heads for the morning cooking and cleaning. In terms of housing materials, only 37 per cent had homes made of brick, while all others were shacks of scrap wood, wattle and daub or a mixture of materials. Now, 97 per cent of the houses are made of brick.

As for household appliances, television ownership went up from 64 per cent to 98 per cent; refrigerators from 58 per cent to 98 per cent; and stereos from 25 per cent to 75 per cent. When you look at the next generation, the consumer profile is surprising. Among the children of the original interviewees, 86 per cent have land or cellular phones (as compared with only 2–3 people in each community who had phones when I first did my study), 79 per cent have washing machines, 71 per cent have VCRs, 30 per cent have microwave ovens (which I suspect is higher than the average in Rio de Janeiro as a whole), more than one in four (26 per cent) own a car and, most surprising of all, 22 per cent have computers. In my interviews, computers were often mentioned as a choice gift for a child's 15th birthday, although parents usually did not want them to have Internet access.

This consumption profile well exceeds that of the average university professor or middle class professional in Rio de Janeiro today. When I give presentations in Rio about this data, the audience is always surprised, but those who work on poverty in other situations in Brazil and Latin America have often expressed to me the recent phenomenon of over-consumption relative to what income levels would suggest. I also notice in the *favela* homes I have been visiting that the kinds of floor and wall tiling in kitchens and bathrooms exceeds in luxury and cost that of most middle class apartments as does the quality of living room, dining room, and bedroom furniture sets. Is this some sort of attempt to rectify the sense of exclusion and stigma that the poor feel as expressed in their constant claim that they are not seen as *gente* (people)?

Educational Mobility

There has also been an extraordinary revolution in literacy and education in general across the generations. Twenty-three per cent of the people interviewed in 1969 were illiterate (mostly among the older members of the population), but that was a huge advance over the 95 per cent illiteracy rate of their parents, who in large part were born and raised in the countryside. In the children's generation, only 6 per cent remained illiterate while 34 per cent went to high school (as opposed to 7 per cent of their parents and zero per cent of their grandparents) and 8 per cent made it all the way to university. Eighteen per cent of the families of our sample have at least one child with a university degree.

The Brazilian school system makes it particularly difficult for the poor to get into universities, which is why this percentage of sample families with university degrees is an impressive achievement. The good high schools are private and expensive, and after completion, there is an expensive preparatory course (*cursinho*) for the critical 'vestibular' exam, which determines who gets admitted to the public universities. These are the best universities, and they are free, but usually only those who have the resources to go to private high schools and are able to pay for the *cursinho* receive sufficiently high grades on the exam to be admitted to public university.

Comparing each original interviewee with his or her parents on a case-by-case basis, 69 per cent had more education than their fathers (their mothers had less). Even more of the children (72 per cent) had more education than their parents, that is, the original interviewees.

Occupational Mobility

Occupational mobility is difficult to measure due to a lack of clear hierarchy among job categories (i.e., the movement of manual to non-manual work may lead to higher or lower salaries and prestige, and 'working for oneself' includes the highest and lowest earners.) However, using the standard categories, we found that the pattern of occupational mobility parallels that in education. With migration to the city, the original interviewees replaced the rural jobs of their parents with unskilled manual labor as measured by 'occupation for the longest period of one's life'. The children of the migrants moved into semi-skilled manual jobs and routine non-manual jobs (see Table 9.5). For example, 62 per cent of the parents of the original sample, 10 per cent of the original interviewees, and none of the children held rural occupations. Comparing each participant specifically with his or her father, 62 per cent of original interviewees (69 per cent of males and 59 per cent of females) had better jobs than their fathers.

Another method of comparing occupational levels is to look at the jobs held during the peak period of the life cycle, which we took to be approximately 20 years after the first job. As jobs tend to vary throughout the life cycle, we compared the jobs of original study participants and their children. It is obvious that the children have done better, but this is more clearly seen on a case-by-case basis: 66 per cent of the children had a better occupation than their own parents; 19 per cent had the same level jobs; and 19 per cent had less prestigious jobs than their parents had. Here is the first indication of a problem with the picture of improving conditions of life over the three generations.

What this reveals is that the impressive educational gains are not fully reflected in occupational gains. Whereas 70 per cent of the original sample had more education than their parents, only 62 per cent ended up with better jobs; and, whereas 75 per cent of the children of the original sample had more education that their parents, only 66 per cent had better jobs. This is, no doubt, an improvement but not as great an improvement as might be expected. Part of this difference is due to a structural change in the value of years of schooling as they relate to prestige in the job market. When I lived in Rio de Janeiro in the late 1960s, parents in the *favela* would often tell their children that if they did not stay in school they would end up as garbage collectors. In July of 2003, the city opened a competition for 400 garbage collector jobs, and 12,000 people applied. A high school diploma was a prerequisite for application.

This change can also be seen in the differential rate of return to education for those living in *favelas* as opposed to the rest of Rio de Janeiro's population. Research by Cardoso *et al.* (2003) has shown that until the third year of school, the increase in income for each year of additional schooling is equal for those in *favelas* and those in other parts of Rio, but from then on, those outside the *favelas* benefit much more.

There is a strong spatial component to the difference in earnings between *favelados* and non-*favelados*. It is much more pronounced in the high-income areas of the South Zone (Leblon, Ipanema and Copacabana) and Barra. These wealthy areas are easy to distinguish on the map of Rio de Janeiro (see Figure 9.5), which has been color-coded by the average income of household heads. The income

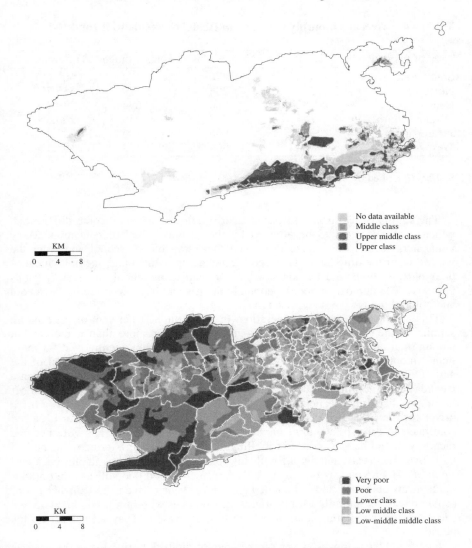

Figure 9.5a and b   Census areas by class of household head

differential between *favelados* and non-*favelados* is greatest in the desirable residential zones, but even in the poorest areas, there is a huge gap. As shown in Table 9.4, the average monthly earnings of non-*favelados* is 567 per cent greater than that of *favelados* in the South zone – even though the South zone *favela* residents earn more than *favelados* in other areas. The smallest difference is in the West Zone, where non-*favelados* earn 153 per cent more per month than *favelados*.

Table 9.4   Average monthly earnings in Reais[9] by residential zone

Area of residence	Favela	Non-favelas	Total average	Difference
South zone	437	2,476	2,173	566.6%
North zone	361	1,284	1,179	355.7%
Near suburb	382	880	694	230.4%
Distant suburb	363	728	655	200.6%
Jacarepaguá	391	896	806	229.2%
West zone	368	564	542	153.3%

Source: Complied by Valeria Pero from the 2000 Census.

The stigma of living in a *favela* is clearly reflected in this income differential, which does not seem to be affected by the 'magic bullet' of education. Many youngsters seem to recognize this when they drop out of school and enter the workforce or the drug trade. They see the underemployment and unemployment of their older siblings and friends who have remained in school and feel that it is pointless. The number of young people, both girls and boys, who are not in school and not looking for work surprised me.

This is worse for the new generation. In 1969, only 31 per cent of the random sample in the *favelas* reported being unemployed for more than a month, but among their children today, 65 per cent say they have been unemployed for more than a month. During many of the interviews I conducted, one or several of the adult children were hanging around the home in the middle of the afternoon on a weekday with nothing to do. The most haunting example is an interview with Mr S., one of the former leaders on Nova Brasília. He had an excellent job as a truck driver for the Coca Cola bottling company on Avenida Itaoca and after it closed continued to drive trucks until his hearing and eyesight failed. His wife became mentally ill and he was reduced to selling off the large, multi-storied home he had built over the years and living with his family in a small hut in the backyard (*quintal*). His home was one of the only houses I visited that still had an outdoor kitchen, an outhouse and a pounded dirt floor. When asked, he said his major problem was getting milk for his grandson, who lives with him, as do his wife and several of their children, none of whom has an income. They all subsist on his pension.

One daughter who was sitting on the edge of the sofa during the interview was in her mid-twenties or early thirties. She did not have a job. She told me she would like to be a supermarket checkout clerk and that she had gone to inquire about a job she had heard was available. When she got to the location, she realized that she lacked enough money for the return fare, so she stayed on the bus and came directly home. The next time she had enough money for the round-trip, but was told she needed a *carteira assinada*, a type of worker document signed by a previous employer. Since she had never worked and did not have the necessary chapters, they told her where to go to register. She went and filled out the forms, and they told her to come back in two weeks. The forms were still not completed, so I asked her how long ago it had been since she had applied, and she looked at her father and said, 'Oh, it's hard to say, about eight months, I think.' That sense of not believing it would make a difference and not even trying anymore was simply not present in the first generation of immigrants. The idea was to do whatever it took to survive in the city.

Another story that sticks in my mind is the young man who had gone to see about a job as bus-fare collector. First of all, he said he found that type of work humiliating; it was ok for his father's generation, but he expected to do better than that. After having been told what the pay would be, he subtracted his travel and lunch costs and the cost of the clothing and shoes he would need to buy and found out that his net earnings would be so low as to be totally insignificant. In addition, he would be traveling 3–4 hours a day and working another 12 hours. So, there he was, 'flying kites' like a little kid, his father told me, and if the father complained, the son would say, 'Don't pressure me or I'll join the drug trade.'

This is indeed a daunting picture. It leads us to the question about barriers to livelihoods.

Barriers to Livelihoods

The primary barriers we have identified to date are many: (1) the dramatic loss of manufacturing in the Rio de Janeiro Metropolitan Area, which has left thousands of blue collar workers unemployed;[10] (2) the consolidation of the physical space of the city and consequent reduction in construction jobs,[11] which had been a mainstay for unskilled and semi-skilled workers in the boom of the 1960s and 1970s; (3) the belt-tightening of the middle-class, which, along with increases in electro-domestic appliances, fast food, and take-out services, has led to a steep reduction in domestic service employment, typically down from live-in maids receiving 'free room and board' plus 5–6 day per week pay to 1–2 days per week, which was the single major female livelihood source in 1968; (4) technological advances that have replaced many labor-intensive jobs with a few high-skilled ones; (5) higher educational standards for job entry due to structural gains in educational levels; (6) the increase in drug-related violence in the *favelas*, which has depressed the value of the rental and sales properties there; (7) the pervasive stigma against *favela* residents reflected in the job market even when the applicant meets all other qualifications for employment. The sources and degree of perceived exclusion are shown in Figure 9.6 based on the percentages of people who responded 'yes' to the question, 'Is there discrimination on the basis of gender, skin color, *favela* residence, manner of dress or "appearance", place of origin, etc.?'

It is interesting that, yet again, living in a *favela* is seen as being more prejudicial than being dark skinned or female. The perception of racism, however, had gone up from 64 per cent in 1969 to 85 per cent in 2001, leading us to wonder whether there is really more prejudice today or just more awareness of it. Many non-profits, particularly Afro-Reggae, emphasize 'roots' and black pride, and, interestingly enough, the same people classified themselves as darker skinned when asked in 2001 than in 1969, more saying they were black, more saying mulatto and fewer saying white than they had in the earlier interview.

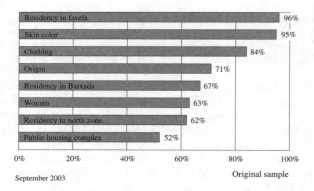

September 2003   Original sample

Figure 9.6   Discrimination as barrier to livelihoods

The issue of mode of dressing came up so frequently in the open-ended interviews that it was included in the final survey instrument even though it had not appeared to be important in the first study.

Dona R., an owner of a clothing and show store in Nova Brasília, who lives in a nearby high-rise and is quite well-off, tells the story of being ignored and then mistreated when she tried to buy a new pair of eyeglasses from an upscale optician in the center of the city. She is a light-skinned woman, perhaps 55 years old, and quite well-spoken. When I asked her why they did not want to serve her, she said it was because she was dressed like someone from the North Zone, not like the South Zone, and they had therefore assumed she was wasting their time as she would be too poor to afford their merchandise.

On the other end of the spectrum, young men evaluate jobs on the basis of how many months it will take them to buy a brand name shirt (or copy) or Nike sneakers, and young women, from maids to manicurists, spend their entire earnings on clothing and accessories that symbolize the South Zone chic. I spoke with one young man who had a cellular phone, a pager, and a palm pilot hanging off the waistband of his baggy shorts. I asked what he needed all of those for? He looked sheepishly at me, and since we have known each other since he was a young boy, he confessed that none of them worked, they were merely accessories.

Fear, Insecurity and Violence

In 1969, people were afraid of their homes and communities being removed by the government. Today they are afraid of dying in the crossfire between drug dealers and police or between rival gangs, the Red Command and the Third Command, in particular. During the first study, only 16 per cent said that violence and crime were the worst things about living in Rio de Janeiro, and today, 60 per cent think so. The fear for personal safety is well-justified. More than one in four people (27 per cent) said that some member of their family had died in a homicide, a level comparable with that during civil wars and much higher than that of cities in the drug-producing countries of Colombia or Bolivia.

One of the most perverse results of the new 'sphere of fear' is the decrease in social capital, one of the few clearly helpful assets in getting out of poverty, or at least relatively improving one's living conditions. People are simply afraid to leave their homes. As N., a former Catacumba resident, now living in Guaporé, put it:

> To live in a place, where you do not have the liberty to act freely, to come and go, to leave your house whenever you want to, to live as any other person who is not in jail. It is imprisoning to think, 'Can I leave now or is it too dangerous?' Why do I have to call someone and say that they shouldn't come here today? It is terrible, it is oppressing. Nobody wants to live like this.

With this comes less use of public space, less socializing among friends and relatives, decreased membership in community organizations, and less networking in general. Thus, news about informal jobs and casual work of all types that was passed easily along the grapevine of connection is now more difficult to come by, and people have turned inward. As shown in Figure 9.7, comparing the participation of the original interviewees in 1969 with that of their children in 2001, the one exception to decreased participation is the Evangelical church. For many women, it is their one opportunity to get out of the house and the one 'leisure' activity they permit themselves. It becomes their only social life. Among young men who have been involved with drug traffic, it is the one avenue of escape.

Another indicator of the changing times, as well as the new isolation caused in part by the entrance of outside dealers and their gangs, is the decline in the sense of community unity. For example, among the people interviewed in 1969, 54 per cent said the community was 'very united' and another 24 per cent said 'fairly united'; whereas among their children, almost none said 'very united' and the majority (55 per cent) said their community 'lacks unity'. This may be due to the fact that *favelas* were always at the high extreme of collective help and mutual aid due to the many battles they fought in common and that only a third of the children (35 per cent) live in *favelas* today, but I think there is more to it than that. I sense that even

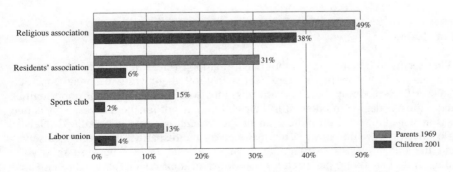

Figure 9.7   Decreased participation in community activities 1969–2001

in the *favelas*, there is much less unity. People feel trapped between the drug dealers and the police and cannot trust either one or their spies. They feel the police do more harm and provide less help than the drug dealers but see both as disrespectful of life in the community. When police enter the *favela* on raids, they barge into people's homes, knock down their doors, knock them around and destroy their possessions, all under the pretext of searching for a hiding gang member. The gang members provoke this by putting a gun to a resident's head and saying, 'Hide me here in your home or I will blow your brains out.'

Lack of Voice

During the period of the dictatorship (1964–85) and especially during the gradual 'opening' and the return to democratic rights, it was assumed that the direct vote for Mayor, Governor and President, which had all been appointed positions, would give the urban poor, a substantial voting bloc, greater bargaining power, a stronger voice with which to negotiate for community improvements and that inevitably this would improve their conditions of life. The interviews did not bear this out. When asked what had changed since the end of the dictatorship, people seemed to agree that there had been improvements in housing, transportation, sanitation, and access to – although not the quality of – education. But, it was also widely agreed that access to health services, personal security and the economic situation, as well as exclusion and bargaining power, had worsened. The *favelados* were deeply disillusioned by democracy's unfulfilled promises. They saw too much corruption and heard too many empty campaign promises, and some had even started to feel nostalgic for the safety and relative peace during the military regime.

Perhaps the election of the first Workers' Party (PT) President, Luiz Inácio 'Lula' da Silva will change this. As of 2001, however, people felt that all levels of government helped them very little and actually did harm in many cases. The state government was seen as the most helpful (still only 37 per cent) and federal government seen as most harmful (52 per cent). Despite all of the funds invested by the Inter-American Development Bank in the massive *Favela-Bairro* Program,[12] only 3 per cent said that international agencies were helpful and 40 per cent gave no response.

Conclusion

What seems to be emerging is the transformation over 35 years from 'the myth of marginality' to 'the reality of marginality'. In 1969, there was widespread hope that the sacrifices made by the city ward migrants, would open up wider opportunities and a greater degree of choice for their children, if not for themselves. This is one reason why the expected radicalism of the *favelados* never materialized (Nelson 1969). They were not angered or frustrated by the disparities between themselves and the upper classes surrounding them in Rio de Janeiro because they were comparing their future perspectives with those of their counterparts who remained in the countryside. Although their children have acquired more consumer goods, have greater access to health care and are more educated, they are suffering from the devaluation of that education in a changed job market, the entrenched stigma associated with living in *favelas* (or *conjuntos*), a shorter life expectancy and a new set of daily dangers involved with the pervasive violence in their communities.[13]

The bottom line for Rio's urban poor is the need for jobs. When asked an open-ended question, 'What is the most important factor for a successful life?' they did not say, better governance, land tenure, or more security from violence. For them it was clear that only a good job with a good salary would get them out of the chronic poverty bind that constrained their freedom (see Figure 9.8).

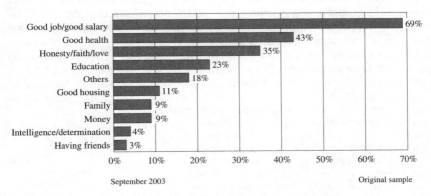

September 2003   Original sample

Figure 9.8   The most important factor for a successful life

The simple fact is that no amount of housing or infrastructure upgrading and no amount of 'integrated community development' or 'partnership strategic planning' can substitute for the primary drive to earn one's living through honest labor. Even the young people who get caught up in drug dealing know they are risking early death and often say they would not be doing these things if there were alternatives.

More years of schooling, often seen as the panacea by policymakers, has not resulted in significantly better access to the changing demands of the job market as shown above. Legalization of land tenure, the great struggle of the 1960s and 1970s, is now a non-issue for *favelados* in Rio de Janeiro, who are no longer threatened with eviction and have *de facto* tenure. In fact, now, that is, decades later, that the literature (see de Soto 2000) and the Cities Alliance policies of the World Bank are focused on land tenure, especially in regard to collateral for loans and the ability to separate capital from assets, Rio's *favelados* are mostly opposed. They do not want to pay property taxes nor submit to building codes, and have no intention of using formal credit systems for loans. People feel it is much too risky to borrow against their homes when they cannot count on a steady income to repay their loans.

Our finding that the most important issue is jobs – whether in the formal or informal sector – and the ability to earn a decent salary, is confirmed by a survey conducted by the Mega-Cities Project in seven of the world's largest cities: London,

New York, Tokyo, Mexico City, New Delhi, Lagos and Rio de Janeiro itself. Our research teams in each city identified the top leaders from the public, private and voluntary sectors as well as academics, labor and media. Across all of these sectors and all of the cities, when asked 'What is the top priority over the next five years?' for that city, the most frequent answer was 'jobs and economic opportunities'.

The one thing that stands out in contrast to the rather dismal situation depicted above is the optimism of the *Cariocas* (Rio de Janeiro's residents). Although the people we interviewed think that conditions in Brazil, Rio de Janeiro, and their own communities may well become worse in the next five years, the majority feel that their own lives will improve, and almost half say that their lives today are better than they had expected. Despite the absence of solutions from the public sector, there are numerous urban social movements, non-profit organizations, and umbrella organizations of residents' associations in *favelas*, *conjuntos* and *loteamentos* that are struggling to turn these signs of optimism into reality.

As I have called my research a Re-Study of Rio's *Favelas*, I would like to end this piece with selected passages of Gilberto Gil's samba *ReFavela*.[14] It captures as only music and poetry can, some of the contradictions and the spirit of *favela* life:

> *A refavela revela aquela que desce o morro e vem transar*
> *O ambiente efervescente de uma cidade a cintilar*
> *A refavela revela o salto que o preto pobre tenta dar*
> *Quando se arranca do seu barraco prum bloco do BNH*
> *A refavela revela a escola de samba paradoxal*
> *Brasileirinho pelo sotaque mas de língua internacional*
> *A refavela revela o sonho de minha alma, meu coração*
> *De minha gente, minha semente, preta Maria, Zé, João*
> *A refavela, a refavela, ó, como é tão bela, como é tão bela, ó*
> *A refavela, alegoria, elegia, alegria e dor*
> *Rico brinquedo de samba-enredo sobre medo, segredo e amor*

> Refavela reveals the one who comes down from the hillside and merges with the
>   effervescent atmosphere of this scintillating city...
> Refavela reveals the leap that the poor black takes when leaving his shack and his
>   community for the cold concrete of the public housing project
> Refavela reveals the paradox of the samba,
> Brazil's specific essence in an international language
> Refavela reveals the dream in my heart and soul
> Of my people, my seed, black Maria, Jose, João
> Refavela, so beautiful, so very beautiful
> Refavela, allegory, elegy, happiness and pain
> The rich and playful source of samba's plot of fear secrets and love

Notes

1   This paper was originally presented as a talk at the Conference on Chronic Poverty in Manchester, England, April 7–9, 2003.
2   Statistics show that family size tends to be larger in *favelas* as compared with the rest of the urban population, despite the fact that *favela* families have been getting steadily

smaller with each generation in the city. For the details of this phenomenon in São Paulo, see Lloyd-Sherlock, Peter (1997).

3   *Bolsa escola* is a small monetary voucher given to the family for each child of school age who stays in school.

4   For more on the concept of advanced marginality see Wacquant (1996). For the ways in which the propositions of advanced marginality fit or do not fit to the reality of Rio de Janeiro's *favelas* see Perlman (2004).

5   When asked who is the politician who has most helped them or people like themselves, the most frequent response was Getúlio Vargas, who during the Estado Novo, between 1930–45, put in place the basis for the welfare state as it is today.

6   We have created a standardized Socio-Economic Status index (SES) using data on family income per capita, years of education, people per room, and consumer goods. The 'mobility index' is a relative index created by measuring the difference in SES scores in 1969 and SES scores in 2001.

7   We will be able to look at these through the analysis of the life history data.

8   This phrase has persisted despite the almost universal paving of *favela* entrances and main passageways through the *Favelas-Bairro* Program, which recently celebrated its 10th year (see note 13).

9   The current exchange rate is US$1 = R$2.7 (*reais*), but it has oscillated a great deal during the past years.

10   During the past decade, 236,078 factory jobs were lost when factories either failed and closed or moved away. One of the people I interviewed in Caxias had been thus displaced by 7 different enterprises over the past 15 years.

11   In the past decade, the Rio de Janeiro Metropolitan Region lost 41,816 jobs in civil construction.

12   The *Favela-Bairro* Program is an upgrading program that aims to improve infrastructure (e.g., streets and sanitation) and social services with community participation. It is a program supported by the Rio Municipality, the Federal Bank (Caixa Economica Federal) and the Inter-American Development Bank, and it originated from a Program of the Urbanization of Popular Settlements of Rio de Janeiro that started as a self-help *favela* initiative. The basic approach of *Favela-Bairro* is to keep the residents in the *favelas*, bring them services available in legitimate neighborhoods (*bairros*) and to improve the physical integration of the *favelas* into the surrounding neighborhoods.

13   According to the New York Times of January 21, 2004, among 60 countries studied by the UN, Brazil has the highest rate of homicide in the world, with 90 per cent by firearms. According to recent studies, Rio de Janeiro follows São Paulo as the city with the highest murder rate, and rates are much higher in *favelas* than in the population at large.

14   Informally translated by the author.

Chapter 10

Unsustainable Trends in Spatial Development in China: Situation Analysis and Exploration of Alternative Development Paths Demonstrated by Case Studies of Kunming (Urban) and Shaxi Valley (Rural)

Jacques P. Feiner and Diego Salmerón

Driving from Kunming International Airport to the city centre, one is taken by all the skyscrapers that have been built in the last 10 years and which now dominate the skyline of this fast-growing city. The huge exhibition hall along the highway is hosting the annual international industrial fair and is decorated with the flags of many nations, symbolizing the foreign presence in this emerging international economic hub. In contrast, a trip to Shaxi, a remote rural commune in the Himalayan foothills about 500 km northwest of Kunming, is depressing in its poverty but gives a good picture of the existing, and growing, socio-economic gap between urban and rural China.

China has re-entered the international stage as a global trading power. Anyone who has visited China's large cities over the past few years has been impressed by the dynamism, the energy, the pace, and the scale of development, whether measured in economic figures, land consumption, construction sites or number of cars. China's economic miracle leaves even sceptics in awe. But this amazing growth has another side. While an increasing number of people in large cities are well-off and have access to modern facilities and information, this development has not yet been translated to the rural areas of China. Thus, remote rural regions in China still live in the 19th century. Though cellular phones have penetrated even the Himalayan backcountry, the daily life of rural Chinese villagers does not differ much from that of 100 years ago. Moreover, between these two ends of the spectrum of fragmented development that has taken place over the last two decades, are smaller urban areas that have been restructured by rural-urban migration and the enforcement of rural industrialization through the establishment of Township and Village Enterprises (TVEs). 'A mosaic of growth and poverty has emerged in China – a regional heterogeneity better understood as the creation of many Chinas,' concluded Muldavin (1998: 290).

Over the last decade, the contrast between urban and rural China, especially in remote areas, has sharpened as China's 'economic miracle' has left the latter behind and created a deepening polarization between a new urban wealthy elite and the poor peasants and farmers. But this contrast is only one of the challenges China is facing today. The rapid pace of urbanization has also accelerated motorization, environmental degradation, and pollution, and has resulted in urban sprawl, loss of farmland, deforestation, erosion, and pollution of air, land and water. The forecast is for population growth and rural-urban migration to continue, thereby increasing urbanization pressure in Chinese towns, cities and mega-cities and paralleled by the depletion of remote rural areas.

This chapter presents an overview of the dynamics of current Chinese development and concludes that one of the paramount challenges is the achievement of sustainable urban development and the stabilization of rural settlements in order to secure China's economic, ecological and social future. Two case studies from the province of Yunnan are presented: the implementation of a comprehensive spatial planning approach in the province capital, Kunming, and the Shaxi Valley Rehabilitation Project, where the goal is to stabilize a remote rural settlement in the Himalayan foothills by initiating ecotourism, accompanied by maintaining the built environment, an improved infrastructure and a diversified economy. Both case studies demonstrate how a comprehensive spatial planning approach can be implemented in a political and administrative environment that is in an incremental transition from a Stalin-style political structure to a market economy.

Current Development Trends in China

Since the beginning of economic reform in 1978, the People's Republic of China (PRC) has experienced dynamic economic development that has penetrated and changed many spheres of life. To give an idea of the strength and impact of these developments, the most important trends are summarized here.

Population Growth

In the second half of the last century, China witnessed steady population growth, despite many initiatives taken since the 1980s to limit it. It grew from an estimated 350 million in 1940 to about 1.3 billion in the year 2000. Forecasts for 2030 can be calculated as follows: Under the current circumstances, the minimum average annual growth rate is 1 per cent,[1] which means China's population would grow to 1.7 billion. Given an unperturbed development, the total population of 2030 cannot be expected to lie below this figure. But, with an average rate increase of only 0.5 per cent (to 1.5 per cent), the population would reach 2 billion,[2] which is a more realistic figure.[3]

Strong Urbanization Tendency

According to official figures, in 1997 only 350 million people (30 per cent) were registered as part of the urban population, while about 70 per cent were registered as

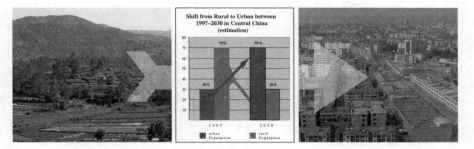

**Figure 10.1   Shift from rural to urban population between 1997 and 2030 in
central China (estimation)**

rural.[4] In the cities, birth control is reputed to be effective, but not in rural areas.
Thus, in 2030 when China's population is between 1.7 and 2 billion, around 80 per
cent[5] or 1.2 to 1.6 billion will be originally of rural descent. Agriculture already does
not offer enough work opportunities for a large share of the rurally registered
population, so they look for work in the cities.[6]

Official estimates say that 640 million[7] people will continue to be economically
dependent on agriculture. This means that between 0.7 and 1.0 billion rurals will
have to look for alternative work and shelter. Looking at present and future
development trends and growth assumptions, we currently expect that by the year
2030, China will reach the urbanization rates of semi-industrialized countries:
approximately 30 per cent rural and 70 per cent urban. If this is the case, between 1
and 1.3 billion rural residents will have to reorganize their lives in an urban way.
When applying even minimal growth rates, and considering the number of people
who will have to move to urban areas, the population that will have to urbanize will
be gigantic. The latter will also require land in order to settle close to where the
economic opportunities are.

Dynamic Economic Growth

While the gross domestic product (GDP) of 1940 was at a minimal level and only
increased gradually up until the 1970s (to about US$ 200 per capita annually), it
rose to about US$ 950 in the year 2000, which is about US$ 1200 billion total GDP.[8]
Thus, in the past 30 years, the annual total GDP (multiplying the growth of the
population with the GDP per capita) has increased more than eight-fold. Official
forecasts predict continuous yearly growth of 7–10 per cent over the next 10 years.
In other words, the total GDP of PRC is expected to double every 10 years.

Industrialization and Tertiarization

Until the late 1970s, agriculture played a dominant role in the economy of the PRC.[9]
Although the first industry's total value has steadily increased, it was largely
surpassed by second and third industry, the latter starting to boom in the 1990s. In

2000, the share of agriculture had already dropped to 15 per cent, while industries took 35 per cent and services 50 per cent of the total GDP.[10]

Rapid Motorization

In 1940, there were almost no motorized vehicles in China. Around 1970, there were approximately 5 vehicles per 1000 inhabitants. By the year 2000, the ratio had already shifted to 40 motorized vehicles per 1000 inhabitants.[11] If this strong motorization trend prevails, we expect 150–250 cars per 1000 inhabitants by the year 2030. If no transportation alternatives are established, there could be between 250–500 million cars in use in the PRC by then. These cars will require fuel, roads, highways, parking space, and will add pollution to the environment as well. Furthermore, car-based transportation is the driving factor for suburbanization and urban sprawl.

Rural Areas Lagging Behind

While strong economic growth of the main metropolitan centers and the coastal regions in general is reported, the interior regions, that is, the central and western provinces,[12] are simultaneously lagging behind. This delay in the development of China's interior regions is even stronger in the rural and mountainous areas than it is in the urban zones. And, when it comes to regions that are populated mainly by the so-called minorities, which mostly live in inaccessible retreat areas, the gap grows even larger. With rising inequality in regional income posing the threat of slowing economic growth and fuelling social conflict,[13] the destabilization of rural settlements through the gradual decay of built structures and the prevalence of floating urban centers with rural-urban migrants, implementing a sustainable economic transformation of rural areas is a paramount task. Even though various scholars[14] have assessed the massive transformation of the countryside and rural systems and policymakers are paying attention to the rural enterprise sector, urban China is still taking precedence – and leaving rural China in the back seat.

Current Planning Policies

During the planned economy period 1949–78, cities were only allowed to expand where the development of heavy industries or administrative constraints made it necessary. After the shift to a social market economy, this policy remained in place and was even adopted in its essence in the respective new laws: the first city planning law (1990) advocated, among other issues, the policy of strictly controlling the scale of large cities. As large cities are not allowed to grow, master plans are very tight. The possibility of a rapid expansion of these cities is thus not even considered.

To prevent the huge rural population (today about 900 million) from migrating into the cities, the *hukou* household registration system was put in place in 1958 (end of socialist transformation phase 1953–57). The *hukou* system had basically two categories: urban and rural. The objective of this registration system was to lock the rural population into their original location and occupation. Until 1978, it was largely effective and kept cities at a minimal level. This policy is still formally in place but was reformed in 2001.[15] But even before that reform, when the new social market economy started to boom with cities as the main centers of attraction, it was no longer possible to keep the huge underemployed rural labor force in the countryside.[16] To offer alternatives to illegal migration into the cities, the *hukou* policy was gradually eased:

- In the first step, villagers were allowed to create new kinds of businesses, so-called Town and Village Enterprises (TVEs). In fact, these mostly low-tech, high energy consuming and polluting small industries have spread all over central China ('Leaving the field, but not the village').
- In a second step, rurals were allowed to move from their villages to the county towns and townships ('Leaving the village, but not the countryside').

Conscious of the limited attractiveness of small towns and townships, the PRC has started to actively promote them by granting planning assistance and other support services, for example, the creation of a new urban infrastructure, such as roads and public facilities. In large cities such as Kunming, enormous investments in a car-based infrastructure have been made during the last few years. This strong promotion of a car-based transportation system has laid the foundation for urban sprawl. The transportation policy focuses more and more on a car-based infrastructure, neglecting mass transit and non-motorized transportation, which would bring a very different settlement pattern.

In the last decade, most of the historic centers were replaced by skyscraper business districts or commodity housing. This physical restructuring of city-centers, which includes a re-location of its original population to the fringes of the city and a destruction of much of the previously existing built heritage, is another typical characteristic of the current planning policy. In general, the main goals of the current policies are:

- Maintaining the status quo, despite on-going strong social (rural-urban) economic (growth) and institutional changes. Schell (2004: 116) points out that China is 'in a state of extreme contradiction, its newly adopted market economy straining against a political structure borrowed from Stalin's Russia.'
- Meeting high short-term economic targets. By focusing on meeting these targets, many long-term sustainability issues related to environmental viability, social equity and economical feasibility are often neglected.

Main Planning Approaches

Up to now, cities and county seats have been carefully planned.[17] They use a master plan, and in their zoning system, the location of housing and industry, public infrastructure and the seats of the government are carefully selected. However, this approach is only focused on the local level of selected urban areas. Meanwhile, the overwhelming majority of the territory, often a densely populated countryside, does not benefit much from the planning. Construction projects such as housing, roads, or factories (e.g., TVEs) do not need to be approved by the various specialized government authorities. The impact of potential projects is not assessed and no efforts are made to coordinate their location. In practice, these projects only need the approval of the Economic Planning Commission and for rural areas, to a wide extent, only the municipal or provincial land administration authority is responsible for land use management. This authority, however, only administrates the land and does not develop concepts for its future use. Thus, there is no spatial planning in its proper sense.

This is further illustrated by the absence of a spatial planning law in the constitutional framework of the PRC. There are actually many laws and regulations that have something to do with spatial planning, but, as these laws and regulations are fragmented, they are not very effective. On the national, provincial and municipal levels, there are administrative units for urban planning, for social, economic and environmental issues, and for infrastructure planning (e.g., water and sewerage, transportation, etc.). What is essentially lacking is a coordination instrument that obligates the different stakeholders to cooperate, at least on a local and regional level.

Spatial planning coordinates space-relevant factors on the local, regional and national levels, and manages the mediation between actors from quite different backgrounds. The importance of coordination rises exponentially with intensified use of the territory. Many provinces in China have extremely high population densities, emerging industries and services, and a developing countryside, but no comprehensive planning approach. In fact, research clearly shows that in the light of the strong development trends in the PRC, the current territorial administrative setup and the spatial planning system is no longer sufficient to meet the challenges of future development.

Consequences of Current Development Trends and Planning Policies

The consequences of restricting the growth of large cities and preventing rural-urban migration will be a lingering, scattered, low-density urbanization (i.e., high per capita land use) of large parts of the PRC's peri-urban countryside[18] and a huge loss of arable land, thus threatening the very basis of their existence. Indeed, if land consumption follows the current trend, land use for urban settlement will increase five-fold by 2030. At the same time, this development trend will also lead to a loss of economic function in many rural and mountainous areas where poverty and economic underdevelopment are already widespread. There one can observe the gradual decay of built structures and the increasing absence of an economically active population, who have left to become part of the so-called 'floating population' in China's large cities.

The consequences of the strong mobilization trend in combination with the high investment in car-based transportation and the neglect of mass transit and non-motorized transportation will be a complete dependence on car-based transportation. Given the vulnerability of large Chinese cities to rapid motorization, frequent urban congestion all the way up to traffic disasters will become the most likely scenario and will demand further high investment in a car-based infrastructure. This will again push urban sprawl. This is a vicious circle that is nearly impossible to reverse once it has started.

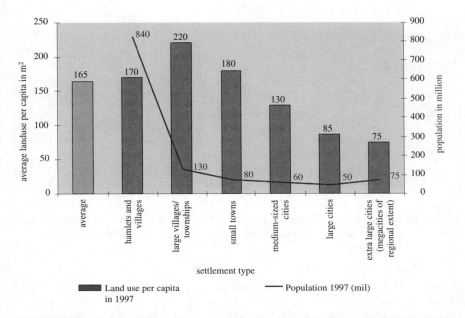

Figure 10.2   Land use per capita and total population in China according to settlement type in 1997

The consequence of favoring small settlements and a highly decentralized industrial structure is tending to be uncontrollable environmental pollution and an increase in pollution-related diseases. As well, noisy concentrations of car-based transportation networks and the lack of green areas will negatively influence many urban and suburban areas and cause them to deteriorate. An additional consequence of neglecting the built heritage and public urban space will be the creation of a banal urban landscape, a loss of urban identity and decreasing life quality for the average urban citizen.

Two Case Studies in Yunnan Province: Kunming and Shaxi Valley

Current situation and problems

Kunming City   Kunming is not only the capital of Yunnan Province, it is also an important economic, cultural and political center of southwest China. Kunming City Proper (KCP), with between 1.5 and 2 million inhabitants, is by far the most important economic motor of development in the province. The Greater Kunming Area (GKA) together with three smaller neighboring industrial cities (Yuxi, Qujing and Chuxiong around 100–150km away) forms the central economic region of Yunnan, and contributes 80 per cent to the provincial GDP.

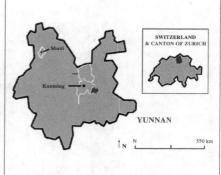

Figure 10.3   Location of Yunnan province, Greater Kunming Area (GKA) and Shaxi in China (size comparison to Switzerland and Zurich canton)

The development trends for the PRC are the same for Kunming. In comparison to the country as a whole, however, we can observe a leveraging effect, which might be typical for most of China's city-regions. Thus, the population here doubled between 1980 and 1997, the economy grew six-fold, and the motorization rate is still increasing 10 per cent every year in Kunming City Proper, leading to a doubling of the vehicle fleet every seven years. It is clear that such dynamics will lead to fundamental changes in the social and urban structure and strongly affect the environment as well, leading to considerable risks, as explained in the following paragraph.

Despite all the controls and restrictive policies that are meant to prevent any expansion, Kunming will continue to grow rapidly because economic forces are becoming increasingly decisive in shaping the settlement pattern, to the detriment of the regulations of the original centrally planned economy. As already outlined, the lack of a comprehensive regional to local planning approach, the overlapping administrative hierarchies and competences and the inappropriate planning scales prevent Kunming from developing a realistic plan for its future regional size. The consequence is that the spatial development cannot really be guided. Thus, different issues and stakeholders cannot be coordinated, synergies cannot be tapped, and main potentials cannot be safeguarded.

This risk is further heightened by the fact that no instruments have been institutionalized so far to coordinate the transportation and settlement patterns, to implement efficient public mass transit, or to take sustainability issues into account. Uncoordinated mono-centric urbanization, similar to what is usually called 'spontaneous' urbanization, threatens to be the result. In order to cope with the ongoing growth of Kunming, ring road after ring road will be added to the city's mono-centric settlement pattern, provoking an uncontrolled spreading out of built-up areas. With increasing distance from the center, land occupation will become fragmented and of low density. Urbanization will occur wherever conditions are favorable, mostly in areas that are flat and well connected to the urban system.

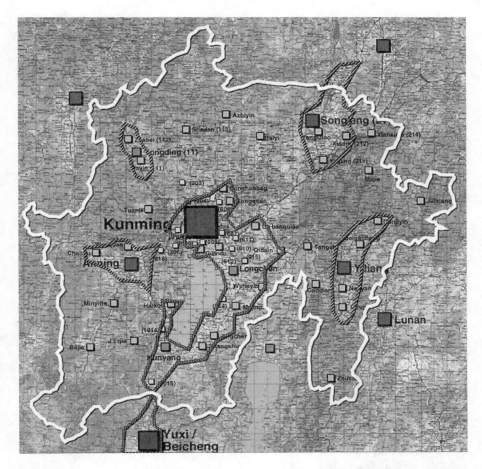

Figure 10.4   Areas with risk of future urbanization in the Greater Kunming Area (GKA)

In this scenario, anything like sustainable development, which means generating a framework for ecological, economic and social issues that balance development and conservation in the long run, will be very hard to achieve. Nevertheless, given the huge risks of the current unfortunate combination of development dynamics and planning policies, we asked ourselves what could be the positive aspects of the current situation. We concluded that Kunming, as with many other medium-sized and large Chinese cities, has just started to develop rapidly and that many of the risks threatening its development had not occurred yet or still had a chance to be reversed. Thus, there is still time to face the threat of urbanization, which is very well illustrated by the following theses:

- Under current circumstances, a huge future population will lead to high urban land use. However, land use is still flexible and can be influenced by strategic planning, saving much fertile land.
- Industrialization and tertiarization is inevitable. A concentration of industrial production locations and the creation of large labor market regions have the potential of significantly improving economic and ecological conditions.
- Future urbanization rates will most likely lead to urban sprawl or low-density large-scale urbanization. However, transportation and settlement patterns can still be structured, reducing investment costs, saving commuting time and urban space, while reducing pollution due to transportation.

Shaxi Valley Driving along the western highway from Kunming to Chuxiong, a medium-sized industrial city in the central economic region of Yunnan province (around 150 km western from Kunming), one is entranced by the magnificent landscapes of this hilly high plateau region. From Chuxiong, a main road continues northwest 200 km to Dali, the capital town of its same named prefecture, situated on the beautiful southern shore of Lake Erhai. From there, local roads continue to lead northwest to the remote areas in the foothills of the Himalayas where our second case study area is situated.

Shaxi, a rural commune in Yunnan Province, is located around 500 km northwest of Kunming, between the tourism cities of Dali to the south and Lijiang to the north. Located in a remote valley in the foothills of the Himalayas and locked into a dead-end situation, it is exactly the kind of rural retreat area mentioned earlier. The

Figure 10.5　The Shaxi Valley
Photo: Jacques P. Feiner.

commune covers an area of 288 sqkm, has eight villages and approximately 20,000 inhabitants, mainly a Sino-Tibetan ethnic group called the Bai, who once dominated large parts of the province.

Due to its position in an interior region and its high altitude level, the valley has remained poor and has not developed much in the last 50 years. In addition, it did not apply for any second industry development. Poverty is thus widespread and jobs are lacking everywhere. Even with natural population growth and the current restrictive migration policies, the population of Shaxi Valley is forecasted – according to official local government figures – to increase considerably in the next few decades, with major parts of its working population migrating temporarily to regional urban centers to find jobs.

On the other hand, the lack of economic growth has resulted in a countryside that is still in very good condition. Much of its cultural heritage has survived, though in other locations this disappeared a long time ago.[19] Importantly for its future, Sideng Market, (previously referred to as Sib Denx) the historic marketplace of Shaxi Valley, originally a way station on the Tea and Horse Caravan Trail between Yunnan and Tibet, has been identified as the last surviving caravanserai-like stopover on this branch of the southern Silk Road. As such, Sideng Market was classified by the World Monument Fund as one of the 100 most endangered world monument sites. But so far, despite its natural beauty and the impressive built heritage, tourism has been limited and has only developed marginally. Insufficient job opportunities and a lack of income drive residents to overuse the natural environment more and more.

Figure 10.6   The Shaxi marketplace
Photo: Jacques P. Feiner.

Thus, the question at hand is how long the current natural beauty of the Shaxi Valley, the state of preservation of its settlements and the unique cultural identity of its inhabitants can prevail under present conditions. In addition, the Bai culture with its own language, its distinct festivals, dances and traditions is nowadays in danger of fading out in favor of the mainstream Han-Chinese culture, which is heavily promoted through all government levels. The same applies to the Bai built heritage, which is often in desperate condition.

Vision and Goals for Future Development in Kunming City and Shaxi Valley

An urban area that is just starting to develop can be more easily guided towards a sustainable pattern than an established urban fabric, which can only be restructured with difficulty. The city of Kunming only started to develop rapidly at the beginning of the 1990s and is still considered to be in the first development stage. Emerging urban regions like the Greater Kunming Area (GKA) should be actively prepared for their future role as economic and social centers. Their regional structure and effectiveness shall be optimized to accommodate a maximum population by minimizing the negative social and environmental impacts. Thus, infrastructures should be planned and coordinated and the use of space-saving means of transportation should be included. The example of Kunming illustrates the potential as well as the risks of future urban development very well and offers a good idea of how a sustainable settlement pattern could be materialized.

The Shaxi Valley Rehabilitation Project is an alternative planning and development project that aims at illustrating how rural communes and their society and economy can be developed in a sustainable way by taking advantage of specific local potential and assets. The challenge of the Shaxi Valley Rehabilitation Project is thus to develop and promote tourism and other industries in the valley as a way of becoming economically viable, while preserving and rehabilitating its cultural heritage, its ecological qualities, and its social structure.

Sustainable Spatial Development Approach

The approach is to encompass the development of the Kunming region as a whole. The regional development component was initiated mainly because the urban dynamics of Kunming City started increasingly to affect the surrounding hinterland. Indeed, the Greater Kunming Area (GKA, total area 9500km^2), which encompasses the most sensitive natural areas, including the Lake Dian watershed region, has just started to be involved in Kunming's urbanization process. The risks of losing huge amounts of prime arable land and greatly increasing environmental pollution were obvious.

GIS-based modeling of future settlement development showed that through the development of a strong regional public transportation system, land use and transportation-related emissions could be reduced by about one-third, for example, using a high-capacity, short-range railway system with close coordination with a decentralized but dense settlement pattern. In addition, the various urban centers in Kunming City Proper would be directly interconnected by this system. This would result in a modified settlement pattern, which would make the various urban centers in the Kunming region easier to reach, and would allow the region as a whole to function more efficiently.

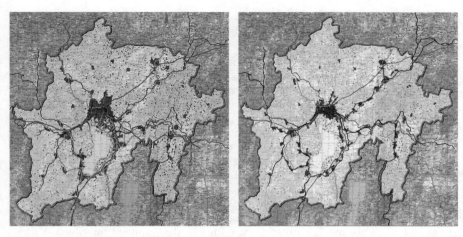

**Figure 10.7   GIS-based regional development scenarios for GKA: trend
(left) and vision (right) settlement and transportation patterns**

To reach this goal, the adoption of the following sustainability-oriented planning policies has been suggested:

- Promoting Kunming as the main future center for population and economic activity in Yunnan Province (and relieving rural areas of surplus population).
- Actively preparing the GKA for its future role as an agglomeration of regional size:
  - Planning and coordinating settlement and transportation patterns.
  - Promoting a decentralized concentration of urban settlements inside the urban region, leaving space for nature areas in between.
  - Promoting mass transportation and non-motorized transportation.
  - Promoting mixed land use around city-centers.
- Promote clustering of same type industries and services at the best adapted locations, and hinder the expansion of polluting low-tech industries in remote rural areas.
- Protect historic sites and city-centers and their original functions by keeping the original population in place.

In addition to the above policies, the implementation of a comprehensive GIS-based planning system for the territorial administrative unit has been suggested to both the city and its region. Therefore, a legal and administrative framework needs to be set up as a base for the implementation of comprehensive planning for the entire administrative unit. In addition, guiding and planning principles for the planning units on the regional, sub-regional and local level have been individually defined and comprehensive, territory-wide planning on all three levels will be implemented step-by-step using cross-sector and multi-disciplinary approaches. The following sub-projects were carried out in detail:

- Suggestions for the reform of the current planning system (implementing comprehensive planning in PRC) were taken up by all the subsequent sub-projects.
- Implementation of GIS-based comprehensive regional, sub-regional and local planning, which will coordinate space-relevant functions and ensure the protection of fertile land and other natural assets.
- Monitoring and controlling methods/approaches for urban and regional development in GKA.
- To put space-related decision-making on a solid base, a Regional Economic Development Strategy has been developed for Kunming. It places the comparative economic strengths of Kunming in a national and international context. As a result, the city will be able to concentrate its funds on its principal strengths and adjust its spatial development accordingly.

Cultural heritage preservation, rural ecotourism and comprehensive planning is suggested as a means for creating sustainable development in Shaxi Valley. Tourism initiated by World Heritage Sites[20] generates revenues and draws global attention to these sites and their cultural assets. At the same time, however, it can cause severe environmental and cultural threats that conflict with the goal of protection and conservation of cultural heritage[21] sites. The close link between cultural heritage and tourism inevitably leads to a discussion of sustainable development, which refers to 'development that meets the needs of the present without compromising the ability of future generations to meet their own needs,'[22] while taking into consideration ecological, economic and social issues. Based on the concept of sustainable development, the World Tourism Organization (WTO) defined the main elements of 'sustainable tourism' in 1995 as follows: '... improvement of the quality of life of the host community ... maintenance of the quality of the environment ... high quality of experience for the visitor.'[23] Though Miller (2001) emphasizes that still no satisfying definition of the concept of sustainable tourism exists, it is commonly agreed that the focus of sustainable (tourism) development has to be on long-term viability as well as the issues of equity, justice, empowerment and participation.[24]

An isolated restoration and tourism project of the Sideng Market in the Shaxi Valley will not be sufficient to preserve the heritage site. Moreover, an integrated and comprehensive planning project, focused on the preservation of the site and the improvement of the economic and ecologic situation in the valley appears to be a prerequisite if the goal is to reach a sustainable and enduring preservation and development. Otherwise, the restored buildings would fall into decay again some years after the program ended and a severe degradation of natural assets would occur. Therefore, the four main objectives of the Shaxi Valley Rehabilitation Project are as follows:

- Enable a sustainable rehabilitation and development of the Shaxi Valley. This should be done by tapping the potential of the valley (i.e., natural and cultural preservation and the development of tourism and other economic potential).
- Enable the development of Sideng Village as the central location of Shaxi commune (i.e., being the core part of future settlement development of the

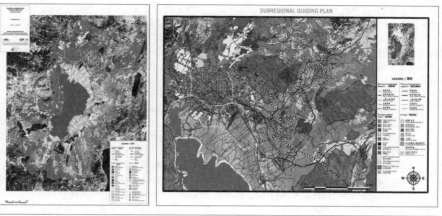

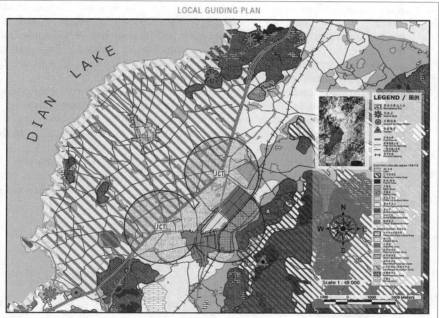

Figure 10.8   Regional, sub-regional and local guide plans for GKA

valley), and the preservation of the historic part of Sideng, as well the other historical assets of Shaxi Valley.

- Ensure the restoration and re-integration of the historic marketplace of Sideng Village.
- Create a model for sustainable rural development in the culturally rich area of the Himalayan foothills.

This model is urgently needed, as many other similar areas are facing similar challenges. Setting an example of how rural sustainable development can take place in such a mountainous region is therefore one of the main objectives. To assure a healthy environment and a sound economic base, a sustainability-oriented comprehensive development plan for Shaxi Commune will be established that includes:

- Comprehensive zoning and transportation plan. This plan addresses, as a novelty for China, the whole area of the commune and outlines core settlement and development zones, nature priority and protection zones, and other areas of common interest.
- Plan for the implementation of a sustainable basic infrastructure. Therefore, a sustainable sanitation system, planned by specialists of the Swiss Federal Institute for Environmental Science and Technology (EAWAG), a Swiss Federal Research Institute, will be established for historic Sideng Village. This system will include liquid and solid waste and its main advantages are that a – very expensive – sewerage system will not be required and that human and animal waste will be used for soil fertilization, at the same time increasing the capacities of organic food production. Sideng Village will be the pilot project for the other villages of the valley. The issues of solid waste, water supply, drainage and grey water are also addressed.
- Tourism development plan. Tourist itineraries, sites of tourist interest and potential locations for hotels and other related infrastructure are outlined.
- Protection and development plan of Shaxi's historic sites. This plan covers the preservation, adequate restoration and re-linking of the historical sites in the valley.
- Creating sustainable investment opportunities in line with the development of the comprehensive development plan, a phased and adapted investment plan for the commune will be worked out. This is needed in order to apply to the

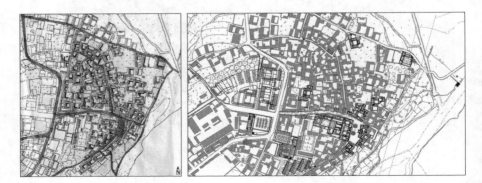

Figure 10.9   Sketch and plan of Sideng Village, showing important buildings and planned interventions on the right
Source: Sketch by Margrit Christensen.

provincial and national levels for funding of infrastructures and to offer a base for private investments. With these funds and investments, which have already been partially promised, the development vision for Shaxi will be realized.

Synthesis and Conclusions

The current government's position that keeping the rural population in villages and townships will be more advantageous for the PRC because the rurals will then have to take care of themselves, is said to be one of the most important motivations for the current policy promoting post-rural settlements and the concomitant TVE industries. But, the analysis of the consequences of the current development dynamics clearly shows that this kind of thinking is outmoded and even threatens the fundamental life resources of the PRC. Moreover, the question to ask is no longer if a large part of the rural population of PRC is getting urbanized, but how it will take place in the context of the PRC. In other words, what kind of (regional) settlement structure should be promoted as the most sustainable one, and how future urban growth could be managed.

For both case studies, the main risk is unsustainable development, but in opposite forms. In Kunming, it is unstructured and scattered large-scale urbanization, which threatens to destroy much development potential. In Shaxi, it is a further loss of spatial and regional functions and decreasing opportunities, which increasingly threaten the main existing potential of the region through decay or overuse. The example of the Greater Kunming Area can be seen as an application of alternative planning policies and comprehensive planning for medium to large Chinese cities. Set up as large city-region with a decentralized concentration of urban centers interconnected by an efficient and environmentally friendly high-capacity, short-range railway system, it has been shown that many of the negative impacts of ongoing development, such as future land use for human settlements, stress on the natural environment, air pollution due to transportation and industrial activity, soil poisoning, urban congestion, costs for settlement and transportation infrastructure and average costs for urban transit can be significantly reduced while improving the economic competitiveness and average life quality. The interpolation of its impact on the PRC suggests that the impact on sustainability could be impressive. Moreover, given the frightening perspectives of the current dynamics, a rapid adoption of alternative policies and the implementation of a comprehensive planning approach designed to prevent large-scale, low-density urbanization and to actively manage development according to sustainability concerns, has become a matter of utmost importance.

In the case of remote rural areas with considerable development potential, such as the Shaxi Valley, we can conclude that though the Shaxi Valley Rehabilitation Project adopts principle ideas of sustainable tourism development in general and rural tourism, as well as ecotourism specifically, its overall philosophy reaches far beyond the limits of tourism development. The comprehensive planning and development approach of the Shaxi Valley Rehabilitation Project is focused on the sustainable development of a rural commune by generating a framework for ecological, economic and social issues that balances development and conservation in a long-term perspective. Tourism will not be the only future for the viability of the Shaxi Valley communities but rather just one of four elements: tourism accompanied by a maintained built environment, an improved infrastructure, and a diversified economy. The overall findings of the Shaxi Valley Rehabilitation Project lead to the conclusion that sustainable tourism development in remote rural areas cannot be achieved without focusing on sustainable community development and establishing a framework for comprehensive planning.

Despite the fact that similar planning approaches have been chosen for the two sites, a generalization of experiences can only be made for city-regions like Kunming; for rural sites like Shaxi Valley, potential experiences depend too much on the local constraints and opportunities. In general, the experiences gathered in completely different cultural and administrative contexts in different situations have been positive. Therefore, we can recommend this as a general approach for many other sites in China.

Notes

1   Tian Xueyuan (1998b) expects a population of 1.4 billion by 2010, which equals a 1 per cent average growth rate (based on the official population figures of 1997).
2   As the one child policy is not really enforced in rural areas (currently 70 per cent of families), the true population growth figure must be higher than the official rate. A 1.5 per cent growth rate would already lead to a total population of 2 billion in 2030.
3   In fact, the Independent Population Research Institute, using a study of grain consumption, reports that the current population in the PRC is actually 1.5 billion. See BBC news under http://news.bbc.co.uk/hi/chinese/china_news/newsid_911000/9113461.stm. In addition, the Japanese Population Research Institution has come to the same conclusion using a study of salt consumption. These research results point to the fact that the problem of population explosion in the PRC is more pressing than usually estimated.
4   The PRC Statistical Yearbook 1999, Beijing: China Statistical Publishing House.
5   Less effective birth control in rural areas will ultimately increase the overall percentage of rural population to 80 per cent by 2030.
6   In 1997, Oskar Weggel (1999) estimated that the so-called floating population was 350–450 million.
7   In Population Problem, it states that Tian Xueyuan (1998a) expects a rural population of 640 million by the year 2050.
8   CIA – The World Factbook 2000.
9   Estimated GDP share of 1970: agriculture 50 per cent, industry 30 per cent, services 20 per cent.
10   CIA – The World Factbook 2000.
11   The PRC Statistical Yearbook 1999, Beijing: China Statistical Publishing House.
12   Hare and West 1999; Xu and Tan 2001; Xu and Zou 2000.
13   Hare and West 1999: 476.
14   Hare and West 1999; Xu and Tan 2001; Seeborg *et al.* 2000.
15   The State Development Planning Commission announced in August 2001 that China was aiming to abolish the *hukou* system over the next five years, and that reform was to be carried out in the rich coastal regions as the first step. The reform of the *hukou* system will help increase employment and establish a unified labor market in rural and urban

areas which will allow the distribution of labor forces according to market demands, the official said. (People's Daily, November 1, 2001, http://www.china.org.cn/english/DO-e/21453.htm).

16 Throughout China, the number of surplus workers in rural areas has already exceeded 150 million and is increasing by five to six million every year, according to official figures. (People's Daily, November 1, 2001, http://www.china.org.cn/english/DO-e/21453.htm).

17 The planning and urban design bureaus of the municipalities usually define the master plan and the master plan area. After approval from the Ministry of Construction and the concerned provincial agencies, this master plan comes into effect.

18 Peri-urban countryside refers to rural regions belonging to the catchment areas of medium, large, and extra large cities in the PRC.

19 Rozelle *et al.* 1997.

20 Currently, 730 properties (563 cultural, 144 natural and 23 mixed properties in 125 countries) are on the World Heritage List, which was established under the terms of The Convention Concerning the Protection of the World Cultural and Natural Heritage adopted in November 1972 at the 17th General Conference of UNESCO.

21 Drost 1996.

22 WCED 1987.

23 WTO 1995.

24 Hall 1998; Ahn *et al.* 2002; Miller 2001.

PART IV
COMING TOGETHER

Chapter 11

Sustaining Cosmopolis: Managing Multicultural Cities

Leonie Sandercock

Arriving and departing travelers at Vancouver International Airport are greeted by a huge bronze sculpture of a boatload of strange, mythical creatures. This 20 feet long, eleven feet wide and 12 feet high masterpiece, *The Spirit of Haida Gwaii*, is by the late Bill Reid, a member of the Haida Gwaii First Nations band from the Pacific Northwest. The canoe has thirteen passengers, spirits or myth creatures from Haida mythology. The bear mother, who is part human, and the bear father sit facing each other at the bow with their two cubs between them. The beaver is paddling menacingly amidships, and behind him is the mysterious intercultural dogfish woman. Shy mouse woman is tucked in the stern. A ferociously playful wolf sinks his fangs into the eagle's wing, and the eagle is attacking the bear's paw. A frog – who symbolizes the ability to cross boundaries between worlds – is partially in, partially out of the canoe. An ancient reluctant conscript paddles stoically. In the center, holding a speaker's staff in his right hand, stands the chief, whose identity – according to the sculptor – is deliberately uncertain. The legendary raven – master of tricks, transformations and multiple identities – steers the motley crew. *The Spirit of Haida Gwaii* is a symbol of the 'strange multiplicity', the astonishing cultural diversity that characterizes 21st century cities and regions.

For me, this sculpture is a powerful metaphor of the contemporary urban condition, in which people hitherto unused to living side by side are thrust together in – what I call – the 'mongrel cities' of the 21st century (Sandercock 2003). Most western nations today are demographically multicultural, and more are likely to become so in the foreseeable future. But how does a demographically multicultural nation become a richly multicultural society? Four years ago, the Canadian federal Privy Council and the Office of Intergovernmental Affairs hosted a special workshop, which concluded that the challenge of integrating immigrant populations was the leading policy challenge for Canada's largest cities. Arguably, this is equally true for the nations comprising the European Union, as well as certain cities in South Africa (most notably Johannesburg), and will become a problem for at least some Asian cities in coming decades.

But I want to make an even bigger – a more global – claim. In this age of globalization, in which so much scholarly attention fetishes the space of flows, time-space compression, the informational city, and the dominance of the global over the local, I will argue that the challenge of social integration remains our most significant human problem in cities and neighborhoods. While the economic and technological forces of globalization have spawned important new literatures on urban and regional economic restructuring, on deindustrialization and reindustrialization, Post-Fordism, and the new informational city, something of at least equal importance, has been relatively ignored. The demographic restructuring, which many cities and regions have been experiencing for several decades now as a result of global human migrations shows no signs of relenting, for reasons that I will suggest shortly. This presents a set of new challenges, layered on persistent older problems of exclusion and segregation. How can we quarrelsome strangers live together, managing our peaceful co-existence in the shared spaces of these multi-ethnic/-racial/-cultural cities and neighborhoods? What kind of a challenge is this?

My intellectual and practical project for the past decade has been two-fold: to provide a better understanding of the emergence of 'cities of difference' in the context of globalization and other social forces, and to reflect on the challenges that these mongrel cities present to the city-building professionals (architects, planners, urban designers), to city-dwellers, to urban governance, and to conventional notions of citizenship. In Vancouver, my home, 51 per cent of residents (as of the 2001 census) are of non-English-speaking backgrounds, a pattern emerging only since the late 1980s, and producing comprehensive changes at the neighborhood level. In Minneapolis, 100,000 Somalis have arrived in the past decade. In Lewiston, Maine, a small town of 30,000, the arrival of one thousand Somali families in the past year prompted the Mayor to send a Fax to the leader of the Somali community, asking him to tell his people to '... stop coming here. We can't cope with you' (Hirabe 2003). In a European Union of 350 million people, foreign minorities are becoming an increasingly significant part of the urban landscape, especially in the larger cities (e.g., 30 per cent in Frankfurt). The question of their acceptance, citizenship rights, and integration has suddenly become central to European politics and is crucial to the ongoing social stability of the region. This issue of social stability, and sustainability, is the subject of this chapter. It is a global problem for cities, but has specific manifestations in different cities and nations.

What policy research has made clear in recent years is the economic and demographic necessity of migrants, in part to do the dirty, dangerous and undesirable work that locals no longer want to do, or where there are simply not enough workers to do these jobs; in part to provide high-end technological skills; and in part to ensure longer term growth in the labor force in order for taxes to support a rapidly aging population (Sassen 2000). As Washington-based Dimitri Papademetriou explains:

> The demographic facts are clear. Because the baby-boom generation has failed to reproduce itself adequately throughout the advanced industrial world, its passage from the economic scene will create working age population voids. At the same time, a retirement age bulge will be created unlike anything we have witnessed in modern times – with the added 'wrinkle' of the aged now living much longer than ever before. This means that much higher old age dependency ratios will follow, whereby the taxes of fewer and fewer workers will have to support ever larger numbers of retirees. These facts suggest that over the next two decades, immigrants will likely be relied upon more and more heavily for many important social and economic purposes. Among them are tending to the needs of relatively affluent first-worlders through their labor, helping to keep retirement and public

health systems afloat through their taxes, and, in many cases, keeping production and consumption systems humming. (Papademetriou 2002: 29–30)

Thus post-colonial history has come full circle, as the world's poor – many of them former 'colonial subjects' – travel to the erstwhile metropolitan nations of empire to settle 'a new frontier' (Sassen 2000: 156) in the midst of the West's prosperous societies, in the middle of 'White Nations' (Hage 1998) that have yet to come to terms with their own colonial and neocolonial pasts. For European nations and cities to reconstitute themselves as multicultural societies is a profound reorientation that surely will take more than one generation. Even for Australia and Canada, officially designated multicultural nations since the 1970s, the realization of richly diverse multicultural polities is still some way off, according to my own recent research in Toronto (Sandercock *et al.* 2004; see also Thompson 2003). For the United States to reconstitute itself as a desegregated society and address the social and economic exclusion of ethnic and other minorities requires the acknowledgement of a different kind of war to be fought than the one on terrorism. These are some of the challenges that need to be addressed by urban researchers today.

The cultural diversity that is the product of this age of global migrations, and that will continue to insert itself as a distinguishing characteristic in cities the world over, is seen by some as a threat, by others as an opportunity. The threats are multiple: psychological, economic, religious, cultural, and now, since 9/11, complicated by threats of terrorism. There is a complicated experiencing of 'fear of the Other', alongside fear of losing one's job, fear of a whole way of life being eroded, fear of change itself. These fears are producing rising levels of anxiety about, and violence against, those who are different, who are seen as not belonging – as 'not my people'. This fear, and the behaviors it provokes (e.g., fortressing, exclusion, violence) is as great a threat to the future stability of cities as the much more talked about global economic forces, and is not unrelated to the reality of terrorism for the United States. The American empire, with its global economic and cultural reach, has produced its own backlash in some poor countries. I don't want to go deeply into the subject of terrorism, but I do want to quote what Cornel West, one of the most eminent African-American scholars, had to say in a powerful speech at Berkeley shortly after 9/11:

> As I sympathise deeply with the lives of the brothers and sisters of all colors in the World Trade Centre, I said to myself, for the first time all Americans will feel unsafe, unprotected, subject to random violence and hated as a people. And that's precisely what it's like to be designated as a nigger in America. (West 2001, quoted in Catterall 2003: 424)

As shocking as that sounds, and it is meant to shock, Professor West goes on to create the term 'niggerization' to draw attention to the exclusion and neglect of one group of citizens within the United States' own national borders. This brutal term might also be applied to describe processes of hollowing out, of alienation and victimization, of malign neglect, of certain sectors of cities and regions around the globe (Catterall 2003). It is precisely that reality, and that danger, that I do want to address here by insisting that the challenge of social integration, ongoing in the United States, is equally a challenge for the world's cities in this age of global migrations. The danger is that cultural difference will further fracture an already fragile urban social fabric, as new demands for rights to the city emerge: rights to a voice, to participation, and to coexistence in and the opportunity to reshape the spaces of the social and built environment. The great opportunity is the creation of cities in which there is genuine acceptance of, connection with, and respect and space for 'the stranger', the possibility of working together on matters of common destiny and forging new hybrid cultures and urban projects and ways of living.

The next section looks at how some cities have begun to address these challenges of social integration in constructive and transformative ways. I want to shift from the global to the local, to our streets and neighborhoods, where these issues are most tangible and most troubling. We will travel from Frankfurt am Main in Germany to Canterbury in Sydney, and then to Vancouver, British Columbia, to see what it takes to 'manage diversity' and integrate strangers. The chapter concludes with thoughts about policy directions for the 'mongrel cities of the 21st century' (Sandercock 2003).

When Strangers become Neighbors

When immigrants with different histories, cultures, and needs arrive in global cities, their presence disrupts taken-for-granted categories of social life and urban space as they struggle to redefine the conditions for belonging in their new society. From the migrants' point of view, there is no social space that beckons them. They must endure a painful process of acquiring a new spatial and social sense of belonging, a new sense of home. From the hosts' point of view, there is an equally painful disruption of their own familiar space, from the imaginary space of the nation to the very real spaces of neighborhoods: the kinds of shops and churches, the smells coming from houses and restaurants, the way people dress, worship, even the way they drive. All this can seem an affront to an established and comfortable way of life. Nostalgia for a disappearing sense of community mixes with fear of and aversion towards the stranger, the outsider. This stranger, on the other hand, has left his/her home – often under circumstances of fear or great hardship – and taken up residence in someone else's 'home', where he or she encounters suspicion, indifference, or even outright hostility. Migrants need to construct a new place that they too can call 'home'.

Migrants have a particularly strong need for community, for practical as well as emotional support, and past experience shows that newcomers will almost always form their own communities, sometimes spatially concentrated (i.e., enclaves), sometimes more dispersed. A truly multicultural society not only encourages and supports community organizations within immigrant groups but also works to incorporate immigrants into wider social and business networks, cross-cultural activities and organizations. How is this second step achieved? How do societies establish civility, then conviviality, across difference? How do 'we' (newcomers and host societies) generate an everyday capacity to live and work with, and alongside, those who are perceived as different?

To become a multicultural society – or a cosmopolis as I have called it –, requires more than a top-down declaration from The Hague, or Berlin, or Washington. And it is more than a matter of bureaucratic management or adjustments to citizenship legislation. It also requires the active construction of new ways of living together, new forms of social and spatial belonging. This is a long-term process of building new communities, during which fears and anxieties cannot be swept under the carpet, but need to be worked through. This is easy to say, but very difficult to do: difficult politically, while xenophobic and anti-Islamic feelings are on the rise the world over; and difficult to implement, precisely because it means dealing with these fears in the host society as well as with the more obvious material needs of immigrants, such as housing, jobs and schooling. So I turn now, briefly, to the experience of one European city that has tried to tackle fears as well as material needs, followed by the experience of one Australian city that has pioneered and mandated multicultural social planning, and one Canadian city that has tried to build a new, intercultural neighborhood where once Anglo-Europeans were dominant. In so doing, I am consciously shifting from the global frame to a micro-sociological study of neighborhoods, with the intent of exploring ways in which global and local forces are co-constitutive.

Frankfurt am Main (Germany)

From 1989 to 1995, Frankfurt am Main, Germany, under the Red-Green coalition city government, embarked on an ambitious social experiment to create a multicultural city in an anti-immigrant society, establishing AMKA (*Amt für Multikulturelle Angelegenheiten* – the Municipal Department of Multicultural Affairs), in the Lord Mayor's office. AMKA's tasks were to work in collaboration with all the agencies of the state to promote the social integration of the city's 30 per cent foreign (non-German) population, and to work directly in the public sphere, to involve itself in a process of *zusammenwachsen* or 'growing together' of all ethnic groups into a peaceful multicultural society, respectful of difference (Friedmann and Lehrer 1997). In AMKA's own estimation, the successful completion of this process could take as long as two or three generations, the result of a long period of mutual learning, mutual adjustment, and continuing (non-violent) conflict. Their political objectives included the following:

- Reducing the German population's fear of 'the Other' and the number of violent acts against foreigners.
- Encouraging public discussion of migration and the limits of social tolerance.
- Working towards the active participation of newcomers in the public affairs of the city.
- Encouraging the cultural activities of each group of foreign residents.
- Offering in-service training for members of the municipal bureaucracy in intercultural communication.
- Forming a culture of everyday life in the context of immigration (Wolf-Almanasreh 1993, quoted in Friedmann and Lehrer 1997: 68–69).

The extraordinary ambition of AMKA and the Red-Green coalition of moving toward the practical utopia of a multicultural society came to an end in 1995 with the election of the Christian Democrats to City Council. It had been a very bold, innovative experiment on European soil, in a country in which the majority had shown little inclination to turn foreigners into Germans and to share citizens' rights with them. But experiments like these have a life beyond their own short span as inspirational examples, which sometimes pave the way for broader changes. What was particularly significant about AMKA is worth summarizing:

- It dealt with multicultural citizenship at the level of the city and everyday life.
- It was committed to a long-term perspective.
- It promoted mutual learning.
- It recognized and tried to address fear of foreigners, and the violence that often accompanies this fear.
- It addressed the culture of the municipal bureaucracy (e.g., police, teachers, judges, planners).
- It saw its main role as educational, oriented to learning and fostering communication.

One conclusion that might be drawn from Frankfurt's experiment is that for a project of migrant integration at the level of the city, there needs to be multiparty support. A second conclusion might be that support from the national state is essential if conditions of becoming a citizen are to change. A third insight has to do with the micro-sociological work that needs to be done street-by-street, neighborhood-by-neighborhood, and across a range of institutions. I particularly want to emphasize the importance of the daily negotiations of difference in the 'banal micro-publics of the city' (Amin 2002), something that my next two stories also reinforce.

Canterbury, Sydney (Australia)

While European cities struggle to shift personal and institutional mindsets not used to thinking of themselves as countries of immigration, we might expect that new world cities such as those in Australia and Canada, nations necessarily founded on immigration, have come to grips with living with difference. In fact, this is far from true, for reasons related to their founding as racialized liberal democracies (see Sandercock 2003: Chapter 4). Nevertheless, as officially multicultural societies, Australian and Canadian governments have been preoccupied with the challenge of accommodating – rather than assimilating – diversity for at least three decades. How far have they come? There is some very good research now available in Australia based on the first national survey of the responsiveness of local governments to cultural diversity (Dunn *et al.* 2001a, 2001b; Thompson *et al.* 1998; Thompson 2003). I draw on this (especially Thompson 2003) in the following brief summary of the efforts of one local municipality (Canterbury) in Sydney. But first I will present some context.

Until the latter third of the 20th century, Australia had a restrictive immigration policy that was known colloquially as the White Australia Policy, meaning that if you were not of European/Caucasian lineage you were not likely to be allowed to settle. That policy unraveled in the late 1960s, and was replaced by an official endorsement of multiculturalism in 1973, with the publication of *A Multi-Cultural Society for the Future* (Grassby 1973). Over the next decade, national government documents advocated a model of cultural pluralism based on principles of social cohesion, cultural identity, equality of opportunity, and full participation in Australian society. Various federal government programs and agencies over the past three decades have been devoted to overseeing the development and implementation of a national multicultural philosophy. For the most part, state and local governments have lagged behind the federal level in their enthusiasm for implementing this new society. Nevertheless, in the most populous state, New South Wales (NSW), legislation is in place to encourage local government to act responsively to its diverse citizenry. The *Charter of Principles for a Culturally Diverse Society* (Ethnic Affairs Commission NSW 1993) has been incorporated into local government legislation and requires local councils to respond – across all departments and services – to cultural diversity. This was done in 1998 with an amendment to the Local Government Act requiring councils to develop a detailed social plan addressing issues related to indigenous communities and migrants from non-English-speaking backgrounds. Such institutional pressures have, in principle, compelled councils to confront the need for reform. In practice, as the research cited above shows, a systematic response to the needs of diverse groups is a long way off. 'Some councils were surprisingly ignorant of the nature of their local diversity, and had failed to identify and abandon discriminatory and iniquitous practices in the provision of services and facilities' (Thompson 2003).

Still, there are some remarkable cases where change has been institutionalized, one of which is in the municipality of Canterbury, 17 kilometers southwest of downtown Sydney, with a population of 132,360 and a growth in the non-English speaking born population from 28 per cent in 1981 to 45 per cent in 1996. Canterbury has seen waves of migration since the 1950s, beginning with Greeks and Lebanese, and more recently, people from China and the Pacific Islands. The Council's Multicultural Social Plan reflects its commitment to cultural diversity, embracing equity and access to quality services for all residents, and the promotion of harmonious and tolerant community relations (Thompson 2003). The Plan defines the settlement needs of all of its diverse residents, after consultation with them, and then identifies actions that link multicultural considerations to all departments within the Council: engineering, corporate and community services, and environmental services. There are 'priority languages' of information and signage, cultural awareness training for staff, and a community worker for multicultural services. There is also a Multicultural Advisory Committee, whose membership is external to the Council, but whose activities are interwoven through all Council departments via their action statements. In these ways, the Council has begun to institutionalize its rhetorical commitments to serve its diverse community.

Under 'Environmental Services' the Council has embarked on several significant initiatives. The Town Centres Development Program focuses on urban design guidelines and reviews the Council's public domain policies for open space, outdoor dining, festivals, temporary outdoor stalls, signage and street furniture. The intention is that public spaces should be well used by all sections of the community, and the understanding is that this can only happen if they are well-designed, with the community's input. There is also a recreation study that draws attention to the need for culturally sensitive recreation policies (from artworks to cultural events to sports facilities to community gardens and landscaping), and a proposed Multicultural Oral History Project that would document the social history of migration into the area, acknowledging the contributions of migrant families to the economic and social life of the municipality.

Thompson (2003) concludes that this and other case studies from the national survey show that it is possible to develop innovative and well-funded projects 'that address cultural diversity as part of mainstream planning activity'. Of course, the fact that Canterbury is one of the most diverse districts in Australia has helped to prioritize this work. And the Council is mandated by state legislation to produce a multicultural social plan. Still, the Council has institutionalized staff positions and sought partnerships with state government to secure funding for projects, and they have used culturally appropriate consultative techniques. What this demonstrates, then, is the desirability of a multi-tiered governmental framework supporting cultural diversity, and of an internal, whole of Council, interdisciplinary approach. At this point, the multicultural project no longer relies on local leadership only or on the vagaries of election cycles, which brought down the Frankfurt experiment.

Vancouver, British Columbia (Canada)

Multiculturalism in Canada has served as a guideline for government policy since 1971. Canadian multiculturalism has encouraged individuals voluntarily to affiliate with the culture and tradition of their choice, and there has been significant spending, through multicultural grants, to support the maintenance of various cultures and languages and to encourage diverse cultural festivals in public places as well as the symbolic gesture of public artworks that recognize and celebrate the multiple peoples who make up the nation. The intention has been to forge a workable national framework of 'unity within diversity', a remarkable change from conventional strategies of nation-building (Mahtani 2002: 70).

Still, there is a significant leap from multicultural rhetoric at the level of national politics and legal frameworks, to what happens in the streets and neighborhoods of Canada's cities and how people relate to each other across cultures. As has been the case in Australia, provincial and local levels of government have been slower to respond to cultural diversity in terms of examining and changing their policies. Recent research in Vancouver (Edgington and Hutton 2002) and Toronto (Wallace and Milroy 1999; Milroy and Wallace 2001) has shown that local policies in relation to the built environment have lagged behind the rapidly changing demographic realities. There are, however, exceptions, beacons of innovation, and it is to one of these that I now turn.

The City of Vancouver (politically, a municipal government within the Greater Vancouver metropolitan region), with a population approaching 700,000, has developed a series of policy responses to its culturally diverse population, including staff hired within the City Planning Department as multicultural planners, and a multicultural outreach program. One remarkable initiative supported by the City is the Collingwood Neighborhood House. Collingwood is a residential neighborhood within the City, home to 42,000 people, only 32 per cent of whom speak English as their native language. In less than 20 years there has been rapid demographic and sociocultural change. In 1986, people of Chinese background comprised 21 per cent of the population and people of English background 51 per cent. By 1996, the area was 44 per cent Chinese, 10 per cent English, with Filipino and South Asian groups growing to 5 per cent and 8 per cent respectively. There are also Italians, Portuguese, Vietnamese, and first Nations Scottish, Irish, and German residents.

In 1985, a group of local volunteers established the Collingwood Neighborhood House (CNH) as a non-profit, non-government organization to provide much needed family and childcare services. The CNH was one of the first institutions in Vancouver to develop a multicultural policy, which is part of what makes its story significant (Dang 2002). More important, though, are the details of how this was done, and the (local and national) circumstances that made it possible.

What CNH does is to develop and provide services according to perceived local needs. But there is more to it than that. First, the organization's real purpose – as reflected in its mission statement – is to build community, and its belief is that that cannot be achieved by providing culturally specific services. The very idea of a 'neighborhood house' implies a place with no cultural affiliation or shared interest other than 'creating a community based on common residency'.[1] Thus the approach to programming is 'intercultural'. This marks a crucial shift in Canada's recent approach to multiculturalism. It requires that policies and programs pay at least as much attention to intergroup co-operation as to funding ethno-specific programs and services – and the CNH pioneered this shift. Second, the services are not seen as merely services meeting a need. They are also seen as providing meeting places where people come together and 'connect by engaging in activities together'. Third, residents are engaged as researchers in the investigation of their own community, which further helps in establishing contacts and building relationships as well as empowering locals to become involved in the decision-making and programming at the Neighborhood House.[2] The CNH also conducts regular antiracism education programs and teaches through its consistent policies and actions that community is built through inclusion rather than through drawing boundaries. This is the daily negotiation of difference in the micro-publics of the city, in everyday activities, that is the most appropriate way to foster intercultural contact and exchange. It is the coming together of strangers on common projects and everyday activities of survival and reproduction of life.

Discussion

There are two directions in which I would like to take the aforementioned stories. One is to reflect on the range of policy responses necessary to address the challenges of global migration and social integration. A second is to look more specifically at the kinds of challenges presented to my field, that of urban policy and planning.

Policy Responses to Global Migration and Social Integration

In the spirit of striving to realize the full potential of Canadian multicultural democracy, and of making Canada a global example of peaceful intercultural coexistence, I recently suggested at least seven requirements (in a session with parliamentarians and their policy advisors in Ottawa). The achievement of a rich multiculturalism, as opposed to a 'shallow' version (Hiebert 2003) – that is, a multiculturalism of food and festivals – depends first and foremost on increased spending over a wide range of multicultural programs, like the ones mentioned already, and not only on settlement services and English as a Second Language, although these are absolutely essential. Overall, this means more federal funding to cities. Second, a rich multiculturalism requires multi-tiered political and policy support systems, from federal through provincial to municipal levels, and extending to the work of Non-Governmental Organizations (NGOs). Third, in addressing the challenges of integration in everyday life, the culture and practices of municipal workers (e.g., police, judges, teachers, planners, and service providers) has to be addressed, through antiracism and diversity training such as that provided through the Hastings Institute in Vancouver, established in the late 1980s by the City to run workshops for public sector workers. Fourth, a rich multiculturalism requires reform and innovation in the realm of social policy, from the most obvious (i.e., language assistance) to the creation of institutions like Collingwood Neighborhood House, support for immigrant organizations, provision of culturally sensitive social services, and so on. A fifth requirement is a better understanding of how urban policies can and should address cultural difference. This includes issues of design, location, and process. For example, if different cultures use public and recreational space differently, then new kinds of public spaces may have to be designed, or old ones redesigned, to accommodate this difference. Space also needs to be made available for the different worshipping practices of immigrant cultures: the building of mosques and temples, for example, has become a source of conflict in many cities. And when cultural conflicts arise over different uses of land and buildings, of private as well as public spaces, planners need to find more communicative, less adversarial ways of resolving these conflicts through participatory mechanisms that give a voice to all those with a stake in the outcome. This in turn requires new skills for planners and architects in cross-cultural practices, which is a challenge to our universities and to educators like myself.

A sixth requirement is the elaboration of new notions of citizenship – multicultural and urban – that are more responsive to newcomers' claims of rights to the city and more encouraging of their political participation at the local level. This involves nothing less than openness on the part of host societies to being redefined in the process of migrant integration and to new notions of a common identity emerging through an always contested notion of the common good and shared destiny of all residents. The seventh requirement is an understanding of and preparedness to work with the emotions that drive these conflicts over integration: emotions within the host society of fear, and attachment to history and memory, as well as the status quo; and for immigrants, the desire for belonging, the fear of exclusion, and violence. When some Vancouverites protested the building of so-called 'Monster Houses' by Chinese newcomers in the late 1980s in certain Anglo neighborhoods, this was what was happening. Urban planners, not trained in negotiation skills or cross-cultural conflict resolution, 'solved' the problem through by-laws that imposed one culture's version of 'appropriate' housing on another. This is not how we move towards an intercultural society. Refusing to acknowledge and deal with these emotions is a recipe for failure in the longer-term project of intercultural co-existence. If multicultural cities are to be socially sustainable, their citizens, city governments, and city-building professions need to work collaboratively on all of these fronts (see Sandercock 2003: Chapter 6).

Challenges to Urban Policy and Planning

Historically, exclusion and marginality have been the constant companions of difference. Think of generations of slaves, indigenous peoples, immigrants and other groups who have been excluded and exploited by 'dominant cultures'. In spite of all the talk, in planning, about working for the 'common good', the reality has all too often been otherwise. As a function of the state, planning is one of many social technologies of power available to ruling elites, and has primarily been used to support the power and privileges of dominant classes and cultures. We have seen this in extreme form in colonial planning (see King 1976, 1990; Rabinow 1989), in planning under apartheid in South Africa (see Mabin and Smit 1997; Watson 2002), and in ethnocratic states today, such as Israel, where a dominant ethnicity imposes its power through the management of space (see Yiftachel 1992, 1996, 2004; Fenster 1999a, 1999b). In only slightly less subtle forms, this use of planning as a technology of power has produced residential segregation by race in the United States (Thomas and Ritzdorf 1997; Martin and Warner 2000), and the attempted exclusion of indigenous people from metropolitan areas and country towns in Australia (Jacobs 1996; Jackson 1998). Is it realistic to imagine planning practices that can reverse or address these historic and systemic inequalities? Where, institutionally, would such practices be located? In state planning agencies, or are they necessarily insurgent practices, located in civil society and social movements?

In addressing these questions, we have to be mindful that planning in western cities takes place in a context of racialized liberal democracies and as yet unresolved 'post-colonial' condition (see Sandercock 2003). Planners have not yet sufficiently analyzed their own role in an ever-present yet invisible cultural politics of difference, a historic role that has reinforced the power of the dominant culture as well as the dominant class. Across Europe, North America, Australia, and New Zealand, that dominant culture has been ethnically white/Caucasian and planning, as a technology of power, has been an implicitly racialized practice. If we accept the claims of hitherto excluded groups to their rights to the city, which I have argued is essential to social stability and longer-term sustainability, then it would seem that the culture of planning is on a collision course with the new cultural politics of difference.

There are four different ways in which diversity challenges existing planning systems, policies and practices. First, the values and norms of the dominant culture are usually embedded in legislative frameworks of planning, by-laws, zoning regulations and so on. The planning system of any city or nation thus unreflectively expresses the norms of the culturally dominant majority. A second way in which recognition of the right to difference presents a challenge to planning practice is that the norms and values of the dominant culture are not only embedded in legislative frameworks, but are also embodied in the attitudes, behavior, and practices of actual, flesh and blood planners. A third challenge concerns situations in which the xenophobia and/or racism within cities and neighborhoods finds its expression or outlet through the planning system, in the form of a dispute over, say, the location of a mosque or Hindu temple, the use of a suburban house as a Buddhist community center, or the retailing practices of Vietnamese or Chinese traders. In such conflicts, it is not the planning system as such that is the problem, but rather the fact that the system becomes an outlet for the deep-seated fears, anxieties or aversions of some residents. How might the planning system, or individual planners, respond in constructive ways? A fourth challenge arises when (western) planners come up against cultural practices incommensurable with their own values.[3]

This is merely a thumbnail sketch of the difficulties – both practical and philosophical – faced by one profession in the transition to multicultural cities that address the right to difference and the right to integration. I am now deliberately raising more questions than I am providing answers, to give some indication of the many directions that research must take, especially a research that is action-/policy-oriented.

What I have aimed to do in this chapter is first, to locate and to humanize globalization – it's about places and people, not just abstract economic forces – and to establish a sense of the very real agency of individual and collective actors, rather than going along with a more dominant view that cities and citizens are victims of global forces. And second, I have argued that the issue of social integration must be foregrounded, that the dangers posed by processes of 'niggerization' apply not only to the United States but to cities, regions, and nations across the globe and affect the social stability and sustainability of the planet itself.

Notes

1 In the early 1990s, over a 1000 residents took part in the planning and construction of the CNH's main facility on Joyce Street. Representatives of various communities, based on age, ability, ethnicity, gender, and socioeconomic circumstances, all left their mark on the design of the facility and the policies that govern its operation (Dang 2002: 89).
2 This approach has been particularly successful in assessing needs among youth and seniors. 'Seniors asking seniors' was one such project, another is an ongoing community youth mapping exercise (Dang 2002: 77).
3 For a nuanced example of such a situation, see Fenster 1999a.

Chapter 12

Towards Gigapolis?
From Urban Growth to
Evolutionable Medium-sized Cities

Marco Keiner

Problems of steering the urban development occur not only in 'mega-cities' but also in medium-sized cities in developing countries. All of them consume more and more resources and overstep local, regional and global carrying capacities. It is clear that the growth of population and urban areas cannot continue. Sustainable development is an answer but the implementation of this concept into policy occurs too late. Urban growth in developing countries is a challenge to the management capabilities of national and local authorities.

How large can cities become? Are there limits to urban growth? Will mega-cities continue to grow until they become ungovernable gigapolises? This has to be avoided. Sustainable, and more precisely, 'evolutionable' development is a must. Early action is needed. Thus, this chapter is devoted to the plight of relatively small but rapidly growing cities in developing countries. Their problems are similar to those of the mega-cities but offer the opportunity for earlier intervention. Meeting the needs for the city's inhabitants of today and to improve the quality-of-life of future generations is of major concern. It is clear that there is no alternative to the path toward comprehensive sustainable urban development. Decentralization, good urban governance, clear visions, improved planning, and participation are essential to achieving sustainable urban development.

Urban Growth without Limits?

The Challenge of Urbanization

The world of tomorrow will be urban. Until today, urbanization rates in the post-industrialized world and in Latin America have passed the 75 per cent mark and will continue to grow. Until 2030, the UN Population Division (2001) expects that up to 81 per cent of Europeans and 85 per cent of North Americans will live in urban areas. And, despite slower growth rates, this tendency will probably continue past 2030. On the other hand, the levels of urbanization were relatively low at the beginning of the new millennium in less developed regions. This means that the potential for future urban growth in developing nations is high. In Africa, the share of population living in cities will rise from 37 per cent in 2000 to an estimated 53 per cent in 2030, and in Asia, the same figures will mount from 48 per cent to 54 per cent during the same period of time. Thus in 2030, 3.8 billion people will live in urban areas in developing countries, compared to 1.4 billion in 1990 (UN Population Division 2001). This means that 80 per cent of global growth of the urban population will take place in the poorer countries of the Tropics and Subtropics, and from 2000 to 2030, the urban population in developing countries will grow by 60 million people a year, effectively doubling in the period from 2000 to 2030.

The reason for this 'Big Bang' is a population explosion linked to tremendous rural-urban migration processes and the striving of people for an improved quality of life in a globalizing world (Keiner *et al.* 2004). In general, in the world's poorer countries, socio-cultural innovations, economic growth, diversification of income opportunities, and new patterns of self-determination of the individual are limited to the big cities. These attract more and more people and, by causing a drain of brain and labor force from the countryside, enlarge the urban-rural gap (Perlman 1993; Keiner and Schmid 2003).

Global urbanization has resulted in so-called 'mega-cities', which are cities with 8 or 10 million or more inhabitants, and which top the interest of many researchers (see, e.g., Yu-ping Chen and Heligman 1994; Rakodi 1997; Bugliarello 1999; Brockerhoff 2000). Today, according to different sources and spatial delimitations (cities, city-regions, agglomerations), there are between 24 and 28 urban giants with more than 8 million inhabitants, with Tokyo being the biggest with more than 25 to 28 million dwellers. Exact figures on Tokyo do not exist and are not necessary, because it no longer makes a difference if a city has 25 or 28 million inhabitants. The number is just too large to imagine or to control.

What is perhaps more interesting than the sheer size of mega-cities is their speed of growth. Cities or urban agglomerations with more than 5 million inhabitants that are to increase their population by more than 50 per cent between 1996 and 2015 are situated mainly in Asia (Mumbai, Dhaka, Karachi, Delhi, Metro Manila, Jakarta, Lahore, Madras and Bangalore). The same is to occur on the Asian-European fringe in Istanbul, and in Africa in Lagos. The latter will more than double its population from 11 million in 1996 to approximately 25 million in 2015 (UN Population Division 2001).

The result of ongoing urban growth is the emergence of metropolises of unprecedented size whose impact on regional, national and global levels is almost unconceivable.

Limits to Urban Growth?

Thus, the question of how large can cities become arises. The aforementioned projections are valid until 2015/2030. But what can be expected after that? Will the metropolization of the Blue Planet continue? Will our children see a city, say Gigapolis, of 50 million and more? Or will the biggest cities be able to stabilize their size in the near future and allow smaller cities to become new million- or mega-cities?

Another time-honored question has to be added. How many people can live on planet Earth (Cohen 1995), that is, in cities at all? Are there limits to urban, economical, environmental or any other kind of growth? Or ultimately, are there any limiting factors beside the Malthusian core principles? More than 200 years ago,

Malthus theorized that humans do not limit their population size voluntarily but are subject to what he called 'preventive checks'. In other words, population reduction tends to be accomplished through the positive checks of famine, disease, poverty and war (Malthus 1798). Before year 2020, for example, there will be 5 billion urban residents who will not be able to feed themselves. They depend on supplies from rural areas. If farmers continue to give up the production of food in order to migrate to the cities, who will then feed the urban dwellers? Malthus points out that catastrophe can be avoided if mankind perceives the warning signs and if the necessary actions are undertaken in time. So, where are the warning signs?

Anti-Malthusians, like Julian Simon, claim that humans adapt to the problems they create by improvements in productivity and efficiency (Simon 1998). According to Simon, the substitutability of resources is infinite and the human population can continue to grow forever. Thus, the key problem is not that resources will run out but that at present, brainpower is not used enough in order to deal with the problems of food security, energy production and so forth.

Bartlett (1997) states that we are already living at our limits. His 'First Law' relating to sustainability points out that 'Population growth and/or growth in the rates of consumption of resources cannot be sustained'. Also, 'Growth in the rate of consumption of a non-renewable resource, such as fossil fuel, causes a dramatic decrease in the life-expectancy of the resource' ('Seventh Law'; Bartlett 1997). His 'Second Law' states 'In a society with a growing population and/or growing rates of consumption of resources, the larger the population, and/or the larger the rates of consumption of resources, the more difficult it will be to transform the society to the condition of sustainability … [Thus] Sustainability requires that the size of the population be less than or equal to the carrying capacity of the ecosystem for the desired standard of living' ('Fifth Law'; Bartlett 1997). Bartlett summarizes, '… in order to move toward a sustainable society, the first and most important effort that must be made is to stop population growth' (Bartlett 1997).

The point of this chapter is not to judge who is right, Malthus and Bartlett, or Simon. Rather, important questions need to be asked: If what they say is true, what must be done? How can we avoid catastrophes, stop population growth, substitute finite resources? Today, for example, the productivity of both rice in India and the corn yield in Brazil is below 3 tons per hectare. Will it be possible to grow 50 or 100 tons of corn per hectare to feed future urban populations, if necessary? Does technology allow another quantum leap toward the next 'green revolution'? If not, how can substitutes be produced, by whom, on what soil and with what kind of investment?

O'Meara (1999) shows examples of how cities can adapt their consumption to realistic needs, produce more own nutrition and energy, and reuse waste more effectively. Also, von Weizsäcker *et al.* (1997) claim that resources like energy and land, for instance, could be used four times more efficiently if human behavior would change. Daly (1994) adds that sustainable development may only be possible if materials are recycled to the maximum degree possible, and if the annual material output of the economy would not grow. But how far can more efficiency and recycling outweigh the future population growth? How can we avoid that bigger populations will not consume more finite resources?

Urban growth not only impacts nutritional and energetic, hydrologic, and atmospheric resources but also land resources and functions of the less densely settled hinterland. The growing cities eat up the land that feeds them. Keywords in this context are urban sprawl, suburbanization, pressure on woodland, depletion of biotopes, and impoverishment of the sink and buffer functions of surrounding rural areas. As McGranahan and Satterthwaite (2003) point out: 'The goal is not sustainable cities but cities that contribute to sustainable development within their boundaries, in the region around them, and globally'. Thus, the governments of urban centers should be concerned about the impact of the city on the region and behave in a sustainable way even if the outlying territories cannot be spared from the effects of the city's waste and pollution.

Focus must be laid upon the carrying capacity of the planet Earth and of its highly urbanized regions. The 'carrying capacity' of our planet is defined as the largest number of any given species (human) that a habitat can support indefinitely. Indefinitely means without damaging the environment, that is, the resource basis (Giampietro *et al.* 1992). Pimentel (1999) claims that already in 1998, the global population exceeded the Earth's carrying capacity. Have we already passed the outer limits to global urban growth? How big can an individual city-region become, and how many of them can be supported in each of the world's regions?

Food availability is not the only criterion. The ability to be governed, or governability, is another. Some authors argue that contemporary institutional and infrastructural inventions would only allow for large contiguous cities to function at maximum populations of around 20 million (e.g., R. Fletcher, quoted in Chase-Dunn and Weeks 2004). Indeed, this mark has been passed by city-regions like Tokyo, Mexico City, and São Paulo. Whereas it does not seem very surprising that institution and infrastructure are overburdened for both of the Latin American cities (Molina and Molina 2002; Santos 1996), one might be astonished to see that even the major metropolitan area of the 'world city' (Friedmann 1986) and 'global city' (Sassen 1991) Tokyo is judged to be beyond governance ability (Takahashi and Sugiura 1996; Honjo 1998).

Big cities become chaotic and tend to grow together. Science-fiction author Isaac Asimov's Trantor is a fictional planet of which the surface is entirely enclosed in artificial domes. It consists of an enormous megalopolis of 40 billion inhabitants. Although we may think that we are still far from becoming Trantor citizens, we are following this trend in many regions of the world as suggests a closer view on a map of artificial light at night.

Is there a critical size of cities beyond which the authorities cannot keep up with the development? Ronald Bleier comments, 'Ironically and tragically, the larger we grow our numbers, the harder it seems to be able to gain consensus on the connection between the growth in human population and the destruction of the environment that sustains us' (Bleier 2004). Due to the huge size and the multitude and complexity of problems involved, achieving sustainable development in the mega-cities of the developing part of the world seems to be a Sisyphean task.

Governability, Density, Vulnerability and Migration

Will mega cities and smaller but still growing cities become more manageable, more livable, environmentally sounder and safer? The answer is obvious. Some are already borderline ungovernable and it is difficult to imagine how they could make

Figure 12.1   The world at night (NASA 2001)

a U-turn from unsustainable to sustainable development. For Bartlett (1997), 'urban growth management' or 'smart growth' are pseudo solutions: 'Whether the growth is smart or dumb, the growth destroys the environment' (ibid.), adding that the term sustainable growth would then be an oxymoron.

What is more important than the cities' size is how they deal with their size. Thus, an issue to highlight is governance capacity. There are large cities that have been run well, have overcome environmental challenges, and are increasingly clean, as there are smaller cities that are poorly managed and have terrible living conditions. Indeed, some have argued that it is neither the size of the city nor the speed of growth that is most important in explaining poor urban environments, but that the main culprit is a lack of good governance (Satterthwaite 1996; Gilbert 1998). Prud'homme (1996) points out that large cities, if well managed, are more productive than smaller cities because the relationship between urban benefits and urban costs is more favorable in larger cities. Thus, 'A small city poorly managed is bad, a large city poorly managed may be as good (or as bad) as a small city well managed, but a large city well managed is definitely best' (Prud'homme 1996).

Big cities are probably more difficult to manage than smaller cities. Hall and Pfeiffer (2000) point out, '... some of the biggest problems occur in relatively small cities'. In fact, not only mega-cities have mega problems. What counts is the governability, that is, the management skills to cope with the problems of growth (Keiner *et al.* 2004). Thus, attention must also be given to smaller cities. In 2000, there were 22 cities with 5 to 10 million people, 370 cities with 1 to 5 million, and 433 cities with 0.5 to 1 million inhabitants. It is estimated that in year 2015 already, there will be worldwide about 564 cities with more than one million inhabitants (UN 1999). The sharpest increase in new million-cities will occur in the less developed regions.

If not only the size of a city but also its governability are important for sustainable urban development, the focus should not only be on mega-cities but also on smaller cities with less effective governance.

Small towns are becoming cities and big cities continue to sprawl. The bigger a city or a city-region, that is, the denser the population, the more stress factors impinge on inhabitants. The result of the demographic extremes of human conglomerates in urban centers is competition among individuals, lower social cohesion, and crime. We might imagine scenes known from movies like *Blade Runner* or *The Day After*.

While they are growing, cities sharpen inner segregation along income and cultural lines, multiply the coordination tasks for planning, and create more urban diversity. Higher densities may propose gains in economical and cultural terms but can become cumbersome in terms of administration, social care, environmental protection, and control.

The larger cities grow, the higher the urban densities become and the more they become vulnerable to natural hazards (earthquakes, volcanism, etc.), accidents, and terrorism (Pelling 2003). Eight of the mega-cities in the developing world border an Ocean. Already today, regular flooding catastrophes destroy settlements and life in countries like Bangladesh. If sea levels rise just a few centimeters due to global climate change, the risks for the crowded populations in those cities will increase, too. Rural areas of food production will suffer as well. Many of the world's big cities are located near fault zones, where damaging earthquakes have taken place in the past (e.g., Jakarta, Mexico City and Teheran). Such cities are at great risk.

What to do? Waiting until humans will be able to settle on the planet Mars (Zubrin 1997) is not the right answer. Countermeasures have to be adopted before it is too late. Safe environments depend on social cohesion and on their affordability. The future sufficiency of resources depends on their accessibility. As resources and money are concentrated in the developed world, poor countries cannot count on importing needed raw materials, goods and nutrition. But nearly 90 per cent of the future population growth will take place in the urban areas of developing countries, where money is scarce. Already today, one third of the people living in developing countries, live in slums or squatter settlements. Over 50 per cent of the global population lives on less than $2 per day (The World Bank Group 2004). Thus, the global welfare divide will increase. Despite the fact that in the era of globalization, all humans are living in the same global village, 'cocooning' – the need to protect oneself from the harsh, unpredictable realities of the outside world – is growing in the developed countries. Will intra-generational solidarity be established one day? Will resources, knowledge, and technology become equally – according to the needs – distributed between North and South? We are far from an ideal world.

A survival strategy for urban dwellers of the South could be to migrate to less burdened, less polluted and less problem charged city-regions in the North. But will the cities in the North really become more open ('PubliCities') for migrants from the disadvantaged parts of the globe? There are a lot of obstacles and fears. Where migrations are possible, problems of integration in multicultural and multiracial societies occur (Sandercock 1997).

Sustainable Urbanization?

Thus, the solutions have to be found not only on a global, but also on a local scale. The challenge for more than half of world's population is to be at the same time *homo urbanus* and *homo sustinens*. The task for urban management is to create urban sustainotopes where the needs of the urbanites are met without compromising the quality of life and resources in the rural areas – today and in the future.

The gap between the city and its hinterland, that is, the impacts of the urban area on the rural areas have to be considered. Cities are no isolated islands. Their impact is at least regional, mostly national, and sometimes global. To become sustainable, the cities have to act on all three levels, as well as for themselves on the local level.

There are several definitions for 'sustainable cities'. The Institute for Sustainable Communities (2004) offers an action-oriented definition as follows:

> Sustainable communities are defined as towns and cities that have taken steps to remain healthy over the long term. Sustainable communities have a strong sense of place. They have a vision that is embraced and actively promoted by all of the key sectors of society, including businesses, disadvantaged groups, environmentalists, civic associations, government agencies, and religious organizations. They are places that build on their assets and dare to be innovative. These communities value healthy ecosystems, use resources efficiently, and actively seek to retain and enhance a locally based economy. There is a pervasive volunteer spirit that is rewarded by concrete results. Partnerships between and among government, the business sector, and non-profit organizations are common. Public debate in these communities is engaging, inclusive, and constructive. Unlike traditional community development approaches, sustainability strategies emphasize: the whole community (instead of just disadvantaged neighborhoods); ecosystem protection; meaningful and broad-based citizen participation; and economic self-reliance.

If we admit the historical comparison to Europe and North America, cities during the Industrial Era were motors of economic development, innovation and interaction (Jacobs 1969, 1984; Perlman *et al.* 1998). Urbanization in developing countries today may also lead to local clusters of entrepreneurial enterprise, so-called 'economies of agglomeration' with significant cost advantages for the private sector and for the supply of public services. For example, in compact cities the efficiency of infrastructure investments is increased.

Southworth (1995), Prud'homme (1996) and Heinrich (2001) summarize the most outstanding aspects of agglomeration economies that benefit from the proximity between producers, suppliers, consumers and workers:

- The size of an urban labor market allows for the availability of an adequate and relatively inexpensive workforce, enhances the division and diversification of labor with new job opportunities, and leads to new skill combinations.
- The access of the firms to a relatively large urban market without long transportation paths, opportunities for specialization and for innovations as well as the ability to react to changes in consumers' demands and potentials for sharing common inputs (warehousing, power, etc.).

In 1989, according to the World Bank, about 60 per cent of the GDP of the developing countries was created in cities, and 80 per cent of future GDP growth is expected to occur there (Perlman *et al.* 1998).

Other related benefits of urbanization that accrue directly to the people include, for example, issues of education, interaction and transfer of know-how (Glaeser 1998). Thus, big cities can be centers of culture and 'social advancement'. These opportunities have to be seized by urban management and planning (Devas and Rakodi 1993).

Sustainable and Evolutionable Cities

Sustainability Critically Revisited

Today, sustainable development is an objective acclaimed by almost all groups of society. More than 170 states signed the Rio Documents, and a huge number of cities is undertaking Local Agenda 21 processes. Has the world become more sustainable during the past decade? The general consent on sustainability seems mainly to rest upon the vagueness of the term. The challenge lies now in the application of the term 'sustainable development' on the global, national, regional, urban and neighborhood level. Thus, the task lies in '... the implementation of initiatives that do not merely pay lip-service to the words but actively do justice to the original concept' (Campbell 2000). Today, 'sustainability' and 'sustainable development' are used inflationary and flexibly by scientists, politicians, industrials and others to suit a variety of wishes and conveniences. Bartlett (1997) examined different reports dealing with these terms. He concludes that the use of 'sustainability' and 'sustainable development' is often careless and contradictory, and that these terms are arbitrary and user-defined and have lost their clear sense – if there ever was one.

Another criticism focusing on the issue of international equity says that 'To think that their present circumstances and their present societal arrangements might be sustained – that is an unsustainable thought for the majority of the world's people' (Marcuse 1998). Thus, sustaining a certain unfavorable state does not only evoke the image of near stagnation or non-development but also does not lead to sustainable development.

As the concept of sustainable development is currently being criticized, one has to ask: 'Are there better alternatives?'. Or, at the minimum, 'Are there appropriate concepts that emphasize the importance of environmental and equity aspects that need to be understood to establish a more workable mainstream view of sustainability in the global and urban context?'.

Alternative Futures: Survivability, Equity and Life Chances

Basically, there are two possible answers. 'Survival development' is a concept that existed before 'sustainability' was coined (Meadows *et al.* 1972). It has been neglected since. Meadows insists that sooner or later a scenario of collapse will be inevitable. For him, it is too optimistic to believe in sustainable development, for it is already too late to achieve this goal. As human behavior cannot be changed without obvious need (war, famine, etc.), the problematic structures, pollution and questions that the present generation are about to leave to its heirs would force the next generations to strive for their sheer survival.

Another approach is to incorporate inter-generational equity and the 'Principle of Good Heritage' into the new concept of 'evolutionability' (Keiner 2004; see below).

Vittorio Hösle (1999), following the thoughts of John Rawles, distinguishes three kinds of equity of distribution between humans: social equity, international equity, and equity between generations. The first two types comprise the problem of distribution between people living today. We are talking about the distribution of resources, finances, knowledge, quality-of-life standards, etc. There are huge gaps not only between developed and developing countries but also between rural and urban areas worldwide. Even inside the metropolis, gentrification leads to a segregation between high-income neighborhoods and poor/informal neighborhoods. 'Equity between generations' means equity between present and future generations. The early advocate of future ethics, philosopher Hans Jonas, coined a moral imperative, stating that human actions of today should leave enough freedom to future generations so that they will also be able to act. 'Act so that the effects of your action are compatible with the permanence of genuine human life', or simply, 'do not compromise the conditions for an indefinite continuation of humanity on earth', or again in a positive light, 'In your present choices, include the future wholeness of Man among the objects of your will' (Jonas 1985). These reflections are close to Kant's categorical imperative, 'Act only on that maxim by which you can at the same time will that it should become a universal law' (Kant 1788).

Economist and philosopher Ralf Dahrendorf (1994) argues that opportunities of living (life chances) contrast to ligatures. Whereas ligatures are established bonds of the individual to society, opportunities are the chances to choose, and the potentials of an individual to make decisions. Human development offers new opportunities for choice and alternative action. The moral appeal that can be derived from this is to ensure that the following urban generations will actually find the preconditions to have at least as many options that we have.

The consumption-oriented living of present-day generations is paid with the misery of yet unborn generations. The futurization of ecological problems means an existential danger not only many years in the future but already for the younger generations of today. In reality, ecologists already lament the 'sustainable destruction' of habitats.

The Principle of 'Good Heritage'

Considering this and also the idea of equity between generations, the author proposes the 'Principle of Good Heritage' (Keiner 2004). This principle is based upon the basic idea that we should leave fewer burdens than we inherited ourselves. The task of today's generation should be to transform its heritage from burden to gain, from limitation to freedom of acting, from barely changeable fate to the ability of achieving sustainability. The next generations should not find comparable, but better living conditions than we have. This is claimed for all residents of *terra*, living in the North or the South, in urban or rural landscapes. Instead of sustaining our burdens and limiting the freedom of our children and grandchildren we should create an environment in which they do not have to be worried about survival but also be able to look ahead and reflect on new opportunities, developments and challenges.

Therefore, we will have to augment the social and economical but first of all, the ecological values and qualities of life, and increase the diversity of natural resources. Instead of just maintaining them for the people who will live on Planet Earth when we are gone, there is a need to explore and harness new resources and to find substitutes for those that are non-renewable.

The augmentation of good heritage by creating new opportunities (life chances) and by reducing burdens is based on the theory that humanity is evolving towards a higher quality of life (Veenhoven 2000) and to higher levels of awareness, technological solutions and social organization.

The Concept of 'Evolutionability'

Thus, the principle of good heritage leads us to the concept of 'evolutionability' (Keiner 2004) that emphasizes the environmental dimension of sustainability. 'Evolutionable development' meets the needs of the present generation and enhances the ability of future generations to achieve well-being by meeting their needs as free as possible of inherited burdens.

In short, the concept of 'evolutionability' is not meant to replace 'sustainability', but to guide sustainable development in the desired direction, that is, the ability of future generations to achieve collective and subjective well-being will not just be 'not-compromised', as stated in the Brundtland Commission's definition of sustainable development (WCED 1987), but expressed in positive terms, will even be improved. Thus, the vision of an 'evolutionable development' is the development towards a society that neither wastes nor destroys its means of existence. The use of raw materials (natural resources) and the stress on ecosystems should not go beyond the carrying capacity so that future generations will be able to live in a reasonably intact environment with enough resources. This will enable them – if their population size does not exceed that of today – to live with the same or even more wealth than we do today. In conclusion, 'evolutionable development' is very close to what 'sustainable development' would be, if the environment would clearly be the key element.

However, this first draft of what 'evolutionable development' is must be expanded. A broad theoretical base for the concept of evolutionability needs to be worked out and concrete examples for implementation and application must be developed.

In the meantime, as implementation always limps behind awareness, a minimum of some immediate measures should be taken in order to cope with the unsustainable development of big and exploding cities in the developing world. First, the opportunities of urbanization outlined above should be seized and strengthened, although without neglecting the challenges that accompany them. The biggest influence on the cities' growth – besides natural population growth and migration – stems from public policy, that is, from appropriate or 'good' urban management.

'Good Governance' for Fast Growing Cities

The future task and main challenge for the management and steering of mega-cities and hyper-expanding medium-sized cities will be to make the turn around from a fast growing urban area with growing problems to a sustainable developing city. The main challenge for sustainable urban development is to establish 'good governance'. Governance is based on the idea of the 'social contract' (Rousseau 1762), in which government and authority are determined through a mutual contract between the authorities and the governed; this contract implies that the governed agree to be ruled only so that their rights, property and happiness are protected by their rulers. Once rulers do not properly uphold the rights, liberty and equality of everyone, the social contract is broken and the governed are free to choose another set of governors or magistrates. Are current urban problems so extreme that we should change the rulers? Would new rulers do a better job?

'Good governance', according to the United Nations Economic and Social Commission for Asia and the Pacific (UNESCAP 2004), has the following characteristics:

> It is participatory, consensus oriented, accountable, transparent, responsive, effective and efficient, equitable and inclusive and follows the rule of law. It assures that corruption is minimized, the views of minorities are taken into account and that the voices of the most vulnerable in society are heard in decision-making. It is also responsive to the present and future needs of society.

For rapidly growing cities in the developing world, even if they are sub-million cities, urban development planning towards sustainability is of top priority. These cities must orient their planning and development management towards the predictable future. In so doing, inevitable problems can be at least reduced, the management of metropolitan areas can be improved, and opportunities for more sustainable future development can be made.

Paths to viable solutions, as described by Hall and Pfeiffer (2000), especially for cities of hyper growth, require a trend reversal based on a sustainability-oriented 'good governance'. Concretely, these authors suggest a number of activities:

- Sparing, respectively efficient use of rare resources;
- Limiting population growth (i.e., dropping birth rates;
- Better and longer school education for women, dropping the illiteracy rate and promoting general education and environmental education;
- Fight against HIV/AIDS and other diseases;
- Intensifying international economic collaboration;
- Adaptation to the actual requirements of industry to attract more foreign direct investment (FDI) and to increase productivity;
- Integration of the informal sector into regular markets of procurement and selling (e.g., allocation of small loans);
- Improvement of professional skills training;
- Participatory and cost-saving upgrade of squatter and slum settlements;
- Improvement, based on the privatization of public services, of water supply and sewage disposal;
- Diminishment of air pollution by lowering the emission values of private owned motor vehicles and by litter re-use, recycling and disposal instead of combustion;
- Promotion of bicycle and (private) bus transport.

This ambitious catalog of propositions is a framework in which the specific prerequisites and development potentials for the national and local (communal) level must be specified.

Our own research (Keiner *et al.* 2004) leads to further recommendations to control the cities' growth and take on challenges that currently face urban management:

- Decentralization, definition of clear competences, avoidance of overlapping functions;
- Improving coordination between public authorities and the private sector;
- Enhancing citizen participation and empowerment;
- Defining the appropriate perimeter of spatial management and planning;
- Elaborating visions for urban sustainability;
- Orienting plans to the principle of sustainability, harmonizing visions, plans and legal instruments;
- Steering urban development through indicator-based monitoring and controlling;
- Networking of cities on the issue of sustainable urban development.

Next, some of these points will be formulated more precisely.

- Decentralization: Who can determine what urban citizenship needs in order to become sustainable? What is a priority and what is not? The question of who decides what is needed and what has to be done varies from country to country. One main aspect is the vertical distribution of power and competences between the national (central) government and its different territorial components, including the cities. Perhaps the biggest tasks in this sense are to cope with the horizontal overlapping of functions and with the high degree of centralization of local decision-making. Decentralization is a must for sustainable development on the local level (Keiner *et al.* 2004). Moreover, urban policy and planning requires a clear distribution of functions and competences.
- Visions: More emphasis has to be laid upon a clearer definition of what is 'sustainable urban development' in the specific context of each city concerned. After fixing the objectives, the way to achieve them should be traced. In order to do so, a comprehensive strategy for sustainable urban development that will be binding on urban planning instruments should be worked out.
- Participation: Sustainable urban development requires citizens' participation (Douglass and Friedmann 1997; Holston 1999) and the democratization of planning decisions (Sandercock 2002). To this end, innovations from the grassroots level have to be promoted (see, e.g., Douglass 1995; Abers 1997, 2001).
- Planning: An approach for offering more future opportunities for living could be made in spatial planning. Spatial planning is the discipline that steers the development of our present and future living space. In many countries, the implementation of sustainable development via spatial planning has been

mandated. The guiding principles of spatial planning should be oriented towards the concept of sustainability and/or evolutionability. Planning instruments should be reshaped in order to create more environmental, economical and societal opportunities. Compacting the cities is a must if their sprawl is to be stopped.

Urban policies are applied through a set of planning instruments and by-laws that range from the regulation of land use to the definition of public and private transportation systems. In order to achieve the desired development as outlined in visions and strategies, the following requirements exist:

- Plans must be oriented toward the objectives of the conceptual framework (i.e., focused on sustainable development).
- Plans and by-laws on urban development should not contradict other plans, strategies and legal regulations.
- Plans should cover the 'right' spatial perimeter, that is, they should not just consider the cities in their administrative boundaries but also their interaction with rural regions should be taken into account.

As urban areas spread and consume land, the agricultural areas must be protected. Indicator based monitoring and controlling systems (Keiner 2002) are appropriate tools to steer the development of urban growth and its impact on rural areas, into the desired direction.

- Networking: Generally speaking, networking among cities aims at putting together knowledge, creating synergies and using resources more efficiently. Networking cities are looking for collaboration with other communities, which have similar tasks that can be done more easily, more cheaply and more successfully if jointly undertaken. Networking does not require a spatial connection, it can be a-territorial, that is, virtual (Internet) and global. Some examples of city networks include the following: Eco-City Network, Smart Communities Network, The Cities Alliance, Citizens Network for Sustainable Development, The International Network for Urban Development, Stockholm Partnerships for Sustainable Cities, etc. (Keiner *et al.* 2004). Atkinson (2002) adds that cooperation between science and urban practice plays a key role in finding solutions for sustainable urban development.

Conclusion

Big cities are on the way to becoming 'gigapolises', if no trend reversal of the current development occurs. Smaller cities will become big cities, too. All of them consume more and more resources and overstep the local, regional and global carrying capacity. It is clear that the uncontrolled growth of population and urban areas cannot continue. Sustainable development is an answer, but the implementation of this concept into policy tends to come late. Also, this concept is too ambiguous. Thus, the concept of 'evolutionable development' tries to clearly emphasize the importance of the environment as bearer of natural resources and absorber of human activities. However, the meaning of the term 'evolutionability' has to be refined. Before this new vision for the urban future of mankind is realized, good governance and improved urban management can avoid that mega-cities become victims of Malthusian scenarios in the near future. Action is needed for the smaller, but rapidly growing cities of developing countries, where the main problems and challenges of humanity – today and tomorrow – occur.

As a result, good governance in mega- and medium-sized cities in the developing world is a must. This includes, among others, the responsible use of resources, decentralization and the improvement of the social living conditions of urban dwellers. Also, the growth of populations and cities has to be stopped with appropriate measures. Finally, visions for sustainable development should be worked out and put into practice by participatory urban planning.

Chapter 13

Urban Planning in the North: Blueprint for the South?

Klaus R. Kunzmann

The transfer of technology, knowledge and experience from the North to the South, from the so-called 'developed countries' to 'developing countries' has been promoted over the decades by international institutions as well as national AID agencies. The rationale behind such efforts was to assist these countries in their endeavor to overcome the negative impacts of underdevelopment. Much has been written about the underlying motives for this generous assistance. The arguments range from political ambitions to export promotion, from ideological reasoning to encouraging world peace, and from health to environmental concerns. Throughout the history of globally practiced development aid, the paradigms of transfer changed – first from infrastructure and water to agricultural development, then from integrated rural to urban and squatter redevelopment and, more recently, from sustainable development to institution building. All this has been backed up by continuously changing development paradigms, reflecting the short-term fashions of academics, publishers and AID bureaucrats.

With this in mind, this chapter will explore the transfer potential of practiced urban planning in the cities of the North to the cities of the South. It will be argued that concepts of urban planning are quite different even in the North. Urban problems, urban development instruments and urban policy differ considerably from North America to Europe, ranging from the Anglo-American tradition of urban planning to traditions in Scandinavia and Germany or Italy and France.

The difficulties and traps of transferring expertise will be discussed, such as the vested interests of the development industry in the North, the ideology and paternalism of Northern planners misusing the cities in the South as testing grounds for development theories and the attitudes of urban bureaucrats in the South and their limited absorptive capacity. The chapter will end with a more positive tone by presenting a few transfer action areas and some principles under which urban planning transfer could make sense.

Urban Planning Here may not be Urban Planning There

The popular divide of the globe in two halves is, of course, not quite helpful when it comes to cities. The North is far from being a homogenous entity as is the South. Los Angeles is as different from New York City as London is from Paris, or Munich from Duisburg. Urban development problems in the South are similarly far from being homogenous, just taking city development in China and in Malawi as two extreme examples – or X'ian and Lagos. Nevertheless this categorization is frequently used to differentiate rich from poor countries, developed from underdeveloped ones. However, when it comes to cities and to urban planning – and this chapter is about urban planning – this oversimplistic differentiation is not very helpful. Urban planning in the United States follows different rules from those in Germany, France or Italy. What may be good for Liverpool may not work in Qingdao, and what Freiburg i. Br. (Germany) does (or does not do) to contain urban development may not be useful for Lilongwe. Best practice transfer from one country to another, as promoted within Europe by the European Commission, is extremely superficial. It may even be less successful if information is transferred, for example, from gradually declining urban regions in Europe to fast growing ones in China. And the transfer of knowledge in urban planning and development from the North to the South is not in the center of transfer efforts.

There is another point that has to be clarified in the beginning: What is urban planning? My understanding of urban planning is the guidance of the spatial development of a settlement, be it a small country town or a metropolis, and such guidance includes proactive intervention into city development processes as well as defensive strategies to minimize the spatial impacts of market forces, both in the interest of the people living or doing business in a place.

By definition urban planning theory is not region specific. In contrast, urban planning and development in practice is. For example, even urban planning in Eastern Germany, 15 years after reunification, still differs from planning for cities in Western Germany, though the established regulatory framework was imposed onto the East almost overnight. Despite the fact that urban planners from the West of the German Federal Republic have been temporally assigned to work in the planning departments in cities of the East, the differences have not disappeared. The country is still divided by a common language.

Only at a very abstract level urban planning can be theorized to such an extent that all local, endogenous conditions become irrelevant, only then professional planning knowledge can be transferred from the North to the South. A Chinese or a Ghanaian urban planner may accept the aims and principles of urban planning – comprehensive, communicative, sustainable, holistic, incremental or what ever the mainstream paradigm of a time suggests – though local planners may follow quite different paths to get at least some of the principles working in the city. Postulating participation, for example, or endogenous development is certainly not controversial, though implementing such principles in the course of urban planning in a region and a particular city may require quite different approaches.

To the regret of many planners in the North, urban planning is not a key commodity interest in the transfer business, mainly because urban planning does not have priority among decision makers who negotiate the contracts between the North and the South. Their priorities are rather focused on infrastructure development, local economic development or basic needs projects, hence, in action areas where early results can be seen and documented and equipment can be sold and received.

The practice of urban planning in the South raises many questions: Does it really make sense to contract adventurous architects from the North for NGOs supported slum upgrading projects in Brazilian *favelas*, while highly qualified architects from Brazil design office buildings in Europe? What conclusions can be drawn from Chinese public sector rituals to invite famous architectural firms of the North to design fancy urban quarters in Shanghai or Beijing, while highly qualified local architects are working undercover as local subcontractors, though in the end they do most of the job? Can it be justified to send young Northern urban planners to medium-sized cities in the periphery of Kenya, while qualified local planners, for obvious reasons, rather prefer to stick to the capital city? Are urban planners from the North just accepted and tolerated as the spearheading bankers who ultimately hold the keys to the implementation funds for urban infrastructure? Or are they welcomed because they do the work local planners do not want to do themselves?

This, in turn, brings me to another quite important aspect of knowledge transfer in urban planning. Obviously, the professional traditions and working conditions under which urban planners try to guide urban development in a city differ from country to country. A planner trained as a fully pledged architect follows different concepts of urban planning than one who received training as a spatial planner in Germany or as a community planner in California, USA. In addition, even architectural training in one cultural environment, for example, in France, differs considerably from a similar education in another cultural environment, for example, in Scandinavia. For a German urban planner, sustainable urban development means resource conserving and ecologically sound planning, This very rigid 'green' concept is not shared by most Anglo-American planners, who rather emphasize the more holistic approach to sustainability that takes into account environmental, economic, social and occasionally cultural dimensions of planning. Finally, a planner working as a well-paid civil servant in the Netherlands may think differently about individual survival in his/her office than an underpaid planner in Nepal. It is quite easy to blame the one and to praise the other in the context of sustainable urban planning, which makes the situation even more confusing.

In addition, a team of urban planners in a city planning department of a city in the South may speak in different planning terms depending on where they received their training, for example, one from the University of Cardiff, the other from the ETH Zurich and a third one in the *Institut Français d'urbanisme*, to name just three totally different urban planning education cultures. In some planning schools they just learn how to read mainstream English language planning literature, in others they learn how to design physical plans according to local planning standards or how to organize community participation.

A final point: Urban planners are occasionally tempted to act as benevolent missionaries. They tend to see their mission in the North as creators of the 'good city' for a 'good society', to protect and conserve the natural environment, to improve the quality of the built environment by enabling underprivileged citizens to participate in decision-making processes and to articulate their needs and expectations. As 'architectural planners' they design ideal cities, as economic planners they promote local entrepreneurs, as socially minded planners they dream of peaceful communicative and cosmopolitan communities, and as environmental planners they strive for ecologically sound garden city neighborhoods. Expatriate northern planners advising ministries or local governments in the South often play the role of missionary, as their ancestors did when bringing Christianity to Africa, America or Asia.

Although this perspective may appear to be too negative, it addresses issues easily overlooked or deliberately pushed aside when transferring knowledge from one city to another, from one urban laboratory to another one in another country. Cities have always been laboratories where planners aim to experiment with the creation of livable life spaces, while hiding behind the backs of politicians.

All this demonstrates some of the difficulties when transferring proactive urban planning and development knowledge from cities in the North to cities in the South, not just ideas, philosophies or warnings.

Urban Planning in the North and in the South

Urban planning issues in the South differ considerably from those in the North. When discussing the differences, two notes have to be made. Generalizations, made for the purpose of clarity, are always unfair to single countries. The North is not as homogenous as it may appear from a southern perspective. Vice versa the same is true for the South. Urban planning cultures in the USA differ from those in France, in the Netherlands or in the Nordic countries, as much as the 'pursuit of happiness' in the USA differs from the French concept of *liberté, fraternité et égalité*. This is similarly true for the South. Depending on local traditions and linkages to former colonial powers, urban planning cultures vary substantially from China to Nigeria, from Thailand to Brazil. Even within the South it makes a difference whether one looks into Africa, Latin America or Asia. Urban planning for a mega-city in China does require another approach than planning for a small town in Burma or Nepal. This is obvious. Nevertheless a few generalizations may help to highlight the major differences and to point to the transfer problems present (see Table 13.1).

The degree to which urban planning is politically accepted and technically implemented differs. Values for nature and mobility, for home ownership or education, differ as well, and the willingness to obey, to control or to pay varies from region to region. As long as urban planning is seen and understood as a mere technical discipline with basic principles of how to produce a physical plan, the transfer problems may be minimal. As soon as it comes to the social and economic aims of urban planning and to the implementation of plans, however, the gap between theory and practice, between wishful thinking and professional reality may widen.

Today, planners in cities in the North deal with problems of demographic stagnation or even decline, with changing values of a four-generations society, with metropolization and growing intra-metropolitan disparities. They have to deal with the spatial implications of a high degree of unemployment in slowly deindustrializing regional economies and are confronted with emerging ethnic economies. They face strong public opposition against flagship projects in socially and ecologically sensitive urban spaces. The politically favored deregulation of the established regulatory system and a weak status as public sector developers further hinder their work.

Table 13.1   Urban planning and development in the North and in the South: A few exemplary comparisons

	North	South
City growth	Low growth or stagnation, some urban containment	High growth, urban sprawl and strip development
Key policy concerns	• Resource conservation • Spatial fragmentation • Ethnic separation • Unemployment • Profiling, events and urban marketing	• Infrastructure development • Slums and squatting • Urban health • Exploding property market
Land use control and law enforcement	Sophisticated and quite efficient system of land use control and law enforcement	Weak land use control and law enforcement due to chronically underpaid independent urban planners
Spatial guidance of urban development	Complex system of public sector instruments to guide urban development	Guidance exclusively by urban infrastructure, political intervention and creative action
Land market	Highly regulated and comparably transparent	Predominantly speculative and not very transparent
Stakeholders	Public institutions, banks, large corporations and civil society	Central government, political populists, local tycoons and international corporations

Planners in the South, in turn, have to cope with spatial implications of enormous rural to urban migration and city growth. Their problem is land use control and law enforcement and the lack of political commitment to any rational development of urban space. All this is aggravated by the lack of financial resources to develop urban infrastructure, by speculative property markets and a politico-administrative environment, where local planning has very little freedom to act independently from national agenda setting power.

Clearly the professional environment for planning in the South is quite different from that of the North. Consequently one of the key competencies of a 'good planner' is the ability to easily grasp local conditions and to apply tools and instruments accordingly and with a high degree of sensitivity when it comes to intervening into the life spaces of people. However, planners cannot easily switch from one world to the other; a chain of bridge builders and interpreters are required in any North-South transfer.

Transfer Traps

There are at least seven traps in which transfer workers can fall, fields that render any transfer of urban planning and development knowledge difficult from one country to the other. It would make sense to clarify such issues first before any systematic efforts to transfer urban planning competence to the South are seriously undertaken. These seven fields are as follows:

- The politico-administrative environment: The politico-administrative environment in a city in the South differs considerably from that in the North. Even within Europe the differences are enormous. A mayor in France has much more power than a mayor in Germany. And the ways decisions are made in a city council in Italy would not quite work in a city in Sweden. Knowing the interests of stakeholders and the mechanisms of decision-making processes in a city is one aspect and being aware of the power of the city within the multi-tiered system of governance in a country is another one. Both are necessary in order to be able to 'inject' foreign expertise into city development processes exactly where it is needed. To acquire this knowledge, build up the required local expertise, and find the entrance points for possible intervention takes much time and requires sensitivity. Ultimately, the learning process is the only valuable outcome of a project.
- Land ownership and land market: Those who own the land in a city influence the decisions about its proper use. Consequently, it does not make any sense to plan if the ownership question and the interests as well as the mechanisms of the property market are neglected. Urban planning followed by infrastructure development can make landowners rich, and that is why they try to influence the way and the degree to which the land is used. In Scandinavian cities, the public sector is still a big landowner and exerts much influence on urban development. If this is not the case, as in the majority of cities in the South, spatial development can only be influenced by infrastructure investment (roads, water and energy provision, etc.), which opens a huge arena for political clienteles and corruption. To control the use of land by regulatory measures usually fails because of the weak system of land use control. This is why the Urban Land Institute in Cambridge, Massachusetts, for example, is focusing on exploring the ways and means to regulate the real estate market in Latin America under given regional socio-political systems. Knowing the local land market arena is a pre-condition for any spatial development efforts, though most urban planners from the North struggle when trying to get acquainted with the mysteries of a local real estate property market in the South, which is rarely documented in an easily accessible way.
- Land use control and plan enforcement: The mechanisms of land use control and plan enforcement are complex in all countries, whether they are highly regulated as in the North or subject to informal negotiations in tribal milieus relying on trust and intricate systems of mutual favors as in the South. Knowing these mechanisms and the vested interests of actors when applying such mechanisms is a precondition for any direct intervention in the urban

fabric. However, the time needed to acquire this knowledge usually exceeds the time span of a project. It is the tacit knowledge of local and regional institutions, which is not easy accessible to foreign experts.

- Financial and tax instruments: What has been said above is similarly true for understanding the rationale behind decisions in the land market. As a rule, such decisions are based on financial deliberations of investors and households, which in turn are often linked to the local tax system and to the varying incentives, disincentives and 'holes of the system' in a country. Location decisions are influenced by the nature of financial considerations and taxation and by the way banks are involved in financing the investment. Even the respective system in the North is intricate. More than once, this has been neglected by urban planners when designing their urban regeneration or town expansion schemes, their communities or green utopias.

- The capacity of local city planning departments: The capacity of city planning departments to absorb external advice is limited. This is similarly true for urban policy sections in national ministries responsible for setting the national framework for urban development. As a rule, the nature of the absorptive capacity is totally misjudged and overrated. Given the competence of the staff, their day-to-day obligations and commitments as well as their limited time budget, it is not easy to transfer knowledge to urban planners in such institutions. As long as the transfer competence provided by imported expert manpower is paid by third parties, for example, a development agency in the North, it is welcomed, although only in a non-committal manner.

- The professional environment of urban planners: Urban planners in city governments of the South work under different conditions. As a rule their salary is low and their forced mobility within the institutions is high. In turn, their status as technical bureaucrats in the public sector is low. Their dependency upon irrational politics is at least as high as in the North, if not even higher. Interwoven into a system of mutual favors, they are constantly balancing commitment to the public sector and to the clan, tribe or party. Depending on where they received their qualification, in the region or abroad, they have quite different visions of desired urban futures and the implementation processes required. Consequently in transfer situations, different planning schools and planning cultures may cause considerable communication difficulties.

- The timing of projects: The ambitions of planners from the North working in the South are very much under the dictate of time. Staff contracts are linked to budget allocations for programs and projects, and budget allocations follow another rationale than urban development projects and programs. Planners are jumping on the running urban train for a few hundred miles, then they leave it, while the train continues its journey. It is the time frame that determines the contract of a transfer agent more than the rationality of the project. Under time pressure, for example, because of the expiration of an expert's contract, decisions are made that otherwise may not have been sought, or local stakeholders wait until the end of a project before doing what they wanted to do without offending the foreign experts.

Urban planners from the North cooperating with planners in the South will have to
be aware of these traps, which they may encounter when carrying out their projects.
The formal cooperation contracts, of course, will not refer to such contextual
conditions, which are difficult to articulate anyway. Moreover, given the financial
interests of the development industry, such transfer constraints are usually neglected
to avoid unnecessary tension before any work has started.

Transfer Industries

When urban planners from the North are recruited for the South, a species of 'super
planners' are being searched by development administrations. Long experience in
the South is required and language capability together with practical experience at
home in leading positions in the public sector. However, as such planners are rarely
found for reasons having much to do with the professional field in the North,
particularly in Europe, where urban planners tend to work preferably as civil
servants in the public sector with inflexible employment schemes and hardly any
incentive to go abroad. Consequently, urban planning expertise is mainly bought
from consulting firms or short-term experts on unpaid leave from the public sector
or academia.

There are many transfer agents, bridge builders and interpreters transferring
urban planning competence from the North to the South. They form a dense network
of urban planners who have an interest in the South. They may have an academic
interest in the specific problems of cities in the South or just have a personal interest
in a particular country resulting from intercultural marriages or family relationships,
they may be curious or adventurous, or just like to function as agents between two
cultures. Presumably the majority of transfer builders perceive a project just as a
field of business activities.

Consequently, the transfer of knowledge from the North to the South is in the
hands of powerful and influential 'transfer industries'. These transfer industries in
the North are a complex network of institutions and individuals who organize the
transfer, or make business by providing the required expertise:

- Banks, such as the World Bank (IBRD), the Asian Development Bank (ADB),
  or in my country, the German Bank of Reconstruction (KfW), give loans to
  urban development projects and make considerable efforts to assess the
  feasibility of the projects to justify the loans;
- The UN Family and all its dependant institutions, with the UNDP as the most
  influential knowledge transfer agency, and HABITAT as the urban planning
  and development arm;
- Public development institutions at the national level, such as ministries for
  over-seas development, with their dependant service agencies providing
  technical assistance to Third World clients;
- NGOs of all sorts, from religious organizations to a plethora of benevolent
  foundations, who pursue their social, ecological or cultural policies by
  promoting development projects in the South or spearhead economic interests
  by exploring regional and local territories;

- Consulting firms who make most of their living from overseas contracts. Only a few consulting firms specialize in urban planning and development. They usually offer such services as a side activity when being contracted for designing and developing urban infrastructure systems;
- Public and private universities who provide education and training for students from the South and academic staff for reconnaissance, feasibility studies or evaluation exercises to legitimize politically desirable projects;
- Publishers from the Anglo-Saxon or Latin realm who produce a plethora of publications addressing urban problems in the South and provide the expertise for transfer knowledge between the two parts of the globe.

The employees in all these institutions have a strong interest in the transfer of knowledge. As a rule, they are well-paid to compensate their work under difficult conditions, and as a side effect they are able to see other parts of the world.

This transfer industry in the North is supported by a similar industry in the South, which over the decades has learned how to play the transfer game. If sustainable development is the fashionable paradigm in the North, the NGOs offer their services accordingly and label all their projects as sustainable ones. If institution building is on the top of agendas, they know how to respond to the changing paradigm, and if training is the soup of the day, they will ask for a bowl.

In some countries, NGOs have replaced the official sector as the main recipient of development aid. In more than one case (e.g., in Nepal) the NGO system has become a refined shadow body of the public sector, with former government employees or their relatives acting as local consultants under the more neutral NGO banner.

Much transfer from the North to the South sustains this complex industry and the academic, ethical and commercial bridge builders.

Promoting Appropriate Knowledge Transfer

This brief account of knowledge transfer in the field of urban planning has been quite skeptical. This does not mean that there are no merits in transferring experience in the field of urban planning and development to the South at all. Indeed, cities in the South can benefit from the experience and failures made in the North. As a rule, the ambitions and visions of urban planners in the North and in the South are very much the same. They both have similar concerns for a socially balanced urban development, ecologically sound sustainable development and for the aesthetical quality of the environment. My concern is the inappropriate and unconscious transfer of values from the North to the South, the negligence of the implementation dimension and the hollow rhetoric of much academic literature on urban development transfer.

Let me suggest five areas where the South can benefit from urban experience in the North and where the North could assist the South in the struggle for better cities and more informed urban governance.

- Monitoring urban change: Observing what is happening in cities in the South with the knowledge of urban development processes in the North, speculating on the causes of urban change, analyzing insufficient governance structures and the implications of untamed mobility or environmental degradation may be the most useful 'service' the North could do for the South, as planners in the South neither have the time nor the money nor the required distance to their day-to-day struggle for better cities to engage in such monitoring tasks. Such outside monitoring, however, would only make sense if it is done together with locals and if the results are easily available in the respective region and in the local working language, which in most cases is not English. Ironically, most Third World research, though advancing the knowledge about urban development in the South, is not available in the Third World. Consequently, ways and means have to promoted that improve the transfer of such knowledge to the South via local collaborators or specific, externally financed affordable journals or web-based site information in local languages. To store the knowledge in the PhD libraries of elitist universities is morally unacceptable.

- Promoting urban planning education in the South: The North has a long tradition in educating students from the South. In fact, many universities in the Anglo-American or French and German realms make much of their profits and academic merits by offering such education, which in turn is financed by university alumni, government fellowships or generous public or private foundations. In most regions of the North the system of higher education in urban planning is linked to regional practice, and it is this regional dimension that forms the curricula of planning education programs. However, after 40 and more years of educating planners from the South in the North, universities in the South have the potential and qualified staff to offer their own education programs at BA and MA levels. And they do so with increasing self-confidence and quality. Their support should have priority over attracting tuition paying graduate students to the North to break the vicious cycle of post-colonial dependency. The promotion of planning education programs could include efforts to diminish the dependency upon Western planning literature, didactic training and supporting local research programs or assistance in building up regional libraries.

- Promoting North-South city partnerships: A traditional means of transferring knowledge and 'best practice' from one city to another has been the city network. The cities of Zurich, Switzerland, and Kunming, China, for example, have practiced such transfer with much enthusiasm. The EU Asia Urbs Programme is promoting such cooperation through externally financed pilot projects that support capacity building, partnership development and good governance. Though such partnership programs have a tendency to become one-time traveling opportunities for scientific and political tourism, they may offer good opportunities for mutual learning beyond paternalistic structures, if the individuals on both sides are well prepared for their transfer task. To this end, it may be useful to identify responsive, patient and bilingual tutors, who act as intercultural moderators in the transfer process and are acquainted with the planning cultures in both milieus. Institutional and personal continuity would favor the formation of personal networks and trust.

- Supporting in-house workshops and seminars: Well-prepared workshops and seminars could be another appropriate means to build up competence and trust. However, budget and time restrictions as well as political considerations on both sides are a few reasons why such knowledge transfer via workshops and in-house seminars will not take place either in the city of the South or in the North. Ideally, such events would be useful to double-check outstanding urban projects or conceptualize longer-term spatial visions. The starting point of partnership efforts has to be a well-prepared list of questions and concerns for which Northern expertise is sought. New information technologies have totally changed the accessibility of information in the North for planners in the South. With a click they can find the required information about, for example, a waterfront development project in London or a checklist for evaluating sustainable urban indicators. Hence in-house workshops and seminars are opportunities for intensive cooperation, rather than simple factual information transfer.

- Encouraging South-South transfer: Theoretically, the transfer of planning knowledge from one country in the South to another one would be the ideal way to benefit from southern experience. For many reasons this does not work. Experience in another southern country does not have the same political prestige and merits as experience gained in the North, hence the decision-making milieu will find many real and pretended counterarguments not to support the obvious knowledge transfer. Similarly, the few global HABITAT conferences are just very fleeting occasions to learn from the neighbors in the South. What could be done is to use the planning schools in the South for this transfer by giving them, on a competitive basis, the necessary funds for fellows, research projects, documentation centers, South-to-South online exchange programs, etc.

There are certainly more ways and means to avoid the traps mentioned above and make the urban planning knowledge of the North more accessible to the planners in the South. It would be helpful if only a minor fraction of the AID money spent for huge urban infrastructure development projects could be used for measures in the five action areas mentioned above.

A Final Word

Urban Planning in the North is no ready-made blueprint for the South. Transferring knowledge from one urban policy environment to another one requires modesty, sensitivity and patience. Modesty to be aware of the very limits of one's own knowledge in managing the complex system of a city in a global context, sensitivity in the day-to-day intercultural communication and patience when it comes to harvest the outcome of transfer efforts. To guide sustainable urban development is a long process, much longer than the life span of individual development projects and the job assignments of transfer experts. Cities in the South will develop the way their societies wish them to grow, whether idealistic planners in the North like it or not, just like cities in the North are not the outcome of rational urban master planning.

References

Abdoul, M. (2002), 'The production of the city and urban informalities: the Borough of Thiaroye-sur-mer in the City of Pikine, Senegal', in Enwezor, O., Basualdo, C., Baueret, U.M. *et al.* (eds), *Under Siege: Four African Cities – Freetown, Johannesburg, Kinshasa, Lagos*, Documenta 11, Platform 4, Ostfildern-Ruit: Hatje Cantz Publishers, pp. 337–358.

Abers, R. (1997), 'Learning democratic practice: distributing government resources through popular participation in Porto Alegre, Brazil', in Douglass, M. and Friedmann, J. (eds), *Cities for Citizens: planning and the rise of civil society in a global age*, Chichester (UK): John Wiley & Sons, pp. 39–65.

Abers, R. (2001), 'Practising radical democracy – Lessons from Brazil', *DISP, 147,* 32–38.

Abeyasekere, S. (1989), *Jakarta: a history*, Singapore, Oxford: Oxford University Press.

Abiodun, J.O. (1997), 'The challenges of growth and development in metropolitan Lagos', in Rakodi, C. (ed.), *The Urban Challenge in Africa: Growth and Management of Its Large Cities*, Tokyo: United Nations University Press, pp. 192–222.

Abu-Lughod, J.L. (1999), *New York, Los Angeles, Chicago: America's Global Cities,* Minnesota: University of Minnesota Press.

Adetula, V.A.O. (2002), 'Welfare associations and the dynamics of city politics in Nigeria: Jos Metropolis as case study', in Enwezor, O., Basualdo, C., Baueret, U.M. *et al.* (eds), *Under Siege: Four African Cities – Freetown, Johannesburg, Kinshasa, Lagos*, Documenta 11, Platform 4, Ostfildern-Ruit: Hatje Cantz Publishers, pp. 359–380.

Aguilar, A.G. (1999), 'Mexico City growth and regional dispersal: the expansion of largest cities and new spatial forms', *Habitat International, 23,* 391–416.

Agyei-Mensah, S. (2002), 'Fertility transition in West Africa', *Journal of African Policy Studies, 12* (2).

Ahn, B.Y., Lee, B.K. and Shafer, C.S. (2002), 'Operationalizing Sustainability in Regional Tourism Planning: An Application of the Limits of Acceptable Change Framework', *Tourism Management, 23* (1), February, 1–15.

Akamatsu, K. (1962), 'Historical pattern of economic growth in developing countries', *The Developing Economies, 1,* 3–25.

Alaedini, P. and Marcotullio, P.J. (2002), 'Urban implications of information technology/new electronics for developing countries', *Journal of Urban Technology, 9,* 89–108.

Altimir, O. (1994), 'Distribución del ingreso e incidencia de la pobreza a lo largo del ajuste', *Revista de la CEPAL, 52,* 7–32.

Amin, A. (2002), *Ethnicity and the Multicultural City. Living with Diversity,* Report for the Department of Transport, Local Government and the Regions. Durham: University of Durham.

APHRC African Population and Health Research Centre (2002), *Population and Health Dynamics in Nairobi's Informal Settlements,* Report of the Nairobi Cross-Sectional Slums Survey 2000, Nairobi: APHRC.

Appadurai, A. (2001), 'Deep democracy: urban governmentality and the horizon of politics', *Environment and Urbanization,* **13** (2), 23–43.

Arriagada, I. (1994), 'Transformaciones del trabajo femenino urbano', *Revista de la CEPAL,* **53,** 91–110.

Atkinson, A. (2002), 'International Co-operation in pursuit of sustainable cities', in Westendorff, D. and Eade, D. (eds), *Development and cities,* Oxford: Oxfam.

Attahi, K. (1999), 'Metropolitan governance in Abidjan', Seminar on Development and Urban Africa, Centre d'Estudis Africans, Barcelona.

Bähr, J. and Wehrhahn, R. (1997), 'Polarization reversal in São Paulo', in van Naerssen, T., Rutten, M. and Zoomers, A. (eds), *Diversity of Development,* Assen: Van Gorcum, pp. 166–179.

Barter, P. (1999), *An International Comparative Perspective on Urban Transport and Urban Form in Pacific Asia: The Challenge of Rapid Motorization in Dense Cities,* Perth: Institute for Sustainability and Technology Policy, Division of Social Sciences, Humanities and Education, Murdoch University.

Bartlett, A.A. (1997), 'Reflections on sustainability, population growth and the environment', *Renewable Resources Journal,* **15** (4), 6–22.

Batley, R. (1996), 'Public-private relationships and performance in service provision', *Urban Studies,* **33,** 723–751.

Becker, C.M. and Grewe, C.D. (1996), 'Cohort-specific rural-urban migration in Africa', *Journal of African Economies,* **5** (2), 228–270.

Bennett, J.W. and Dahlbert, K.A. (1990), 'Institutions, social organization and cultural values', in Turner, B.L. *et al.* (eds), *The Earth as Transformed by Human Action, Global and Regional Changes in the Biosphere over the last 300 Years,* Cambridge: Cambridge University Press, pp. 69–86.

Berg-Schlosser, D. (2003), 'Multi-level analyses', in Berg-Schlosser, D. and Kersting, N. (eds), *Poverty and Democracy – Self-Help and Political Participation in Third World Cities,* London: Zed, pp. 181–195.

Bernard, M. and Ravenhill, J. (1995), 'Beyond product cycles and flying geese: regionalization, hierarchy and the industrialization of East Asia', *World Politics,* **47,** 171–209.

Berry, B.J.L. (1997), 'Long waves and geography in the 21st century', *Futures,* **29,** 301–310.

Bleier, R., *Rationale and core principles,* http://desip.igc.org/malthus/principles.html, retrieved on February 10, 2004.

Body-Gendrot, S. (1999), *Economic Globalization and Urban Unrest,* London: Blackwell.

Bogotá, Veeduría Distrital (2002), *Vulnerabilidad social en Bogotá DC 2002, Vivir en Bogotá: Indicadores Sociales,* November.

Bookchin, M. (1994), 'The Population Myth', in Bookchin, M. (ed.), *Which way for the ecology movement?* Edinburgh, Scotland: A.K. Press, pp. 30–48.

Borchert, J.G. and van Ginkel J.A. (1979), 'Die Randstad Holland in der niederländischen Raumordnung', Kiel: Verlag Ferdinand Hirt.

Brasília (Journal of Companhia Urbanizadora da Nova Capital do Brasil – Novacap.

Brautigam, D. (1997), 'Substituting the state: institutions and industrial development in Eastern Nigeria', *World Development,* **25** (7), 1063–1080.

Breheny, M. and Rookwood, R. (1993), 'Planning the sustainable city region', in Blowers, A. (ed), *Planning for a sustainable environment,* London: Earthscan, pp. 150–189.

Brockerhoff, M.P. (2000), 'An urbanizing world', *Population Bulletin,* **55** (3), September 2000.

Brook, T. and Frolic, B.M. (eds) (1997), *Civil Society in China,* Armonk, N.Y.: M.E. Sharpe.

Brunn, S.D. and Williams, J.F. (1983), *Cities of the World: World Regional and Urban Development,* New York: Harper and Row Publishers.

Bryceson, D.F. and Jamal, V. (eds) (1997), *Farewell to Farms,* Aldershot: Ashgate.

Bugliarello, G. (1999), 'Megacities and the developing world', *The Bridge,* **29** (4).

Bulmer-Thomas, V. (ed.) (1997), *The new economic model in Latin America and its Impact on Income Distribution and Poverty,* Basingstoke: Macmillan/New York: St Martin's Press, in association with Institute of Latin American Studies, London.

Burdick, J. (1992), 'Rethinking the Study of Social Movements: The Case of Christian Base Communities in Urban Brazil', in Escobar, A. and Alvarez, S.E. (eds), *The Making of Social Movements in Latin America,* Boulder: Westview Press, pp. 171–184.

Burgess, R., Carmona, M. and Kolstee, T. (eds) (1997), *The Challenge of Sustainable Cities: Neoliberalism and Urban Strategies in Developing Countries,* London: Zed Books.

Burke, G.L. (1966), *Greenheart Metropolis: Planning the western Netherlands,* London: Macmillan.

Caldeira, T.P.R. (1984), *A Política dos Outros: O Cotidiano dos Moradores da Periferia e o que Pensam do Poder e dos Poderosos,* São Paulo: Brasiliense.

Caldeira, T.P.R. (2000), *City of Walls: Crime, Segregation, and Citizenship in São Paulo,* Berkeley: University of California Press.

Caldeira, T.P.R. (2001), 'From Modernism to "Neo-Liberalism" in São Paulo: Reconfiguring the City and Its Citizens', paper presented to the Sawyer Seminar Globalizing City Cultures and Urban Imaginaries, Columbia University.

Câmara dos Deputados, Secretaria Especial de Desenvolvimento Urbano da Presidência da República, Caixa Econômica Federal and Instituto Pólis (2001), *Estatuto da Cidade – Guia para Implementação pelos Municípios e Cidadãos,* Brasília: Câmara dos Deputados.

Campbell, H. (2000), 'Sustainable Development – Can the vision be realized?', *Planning Theory & Practice,* **1** (2), 259–284.

Campbell, T. (2003), *The Quiet Revolution. Decentralization and the rise of political participation in Latin American cities,* Pittsburg: University of Pittsburgh Press.

Capello, R., Nijkamp, P. and Pepping, G. (1999), *Sustainable Cities and Energy Policies,* Heidelberg: Springer-Verlag.

Carter, H. (1983), *An Introduction to Urban Historical Geography,* London: Edward Arnold.

Cartier, C. (2001), '"Zone fever", the arable land debate and real estate speculation: China's evolving land use regime and its geographical contradictions', *Journal of Contemporary China,* **10** (28), 445–469.

Castells, M. (1983), *The City and the Grassroots. A Cross-Cultural Theory of Urban Social Movements.* Berkeley: University of California Press.

Castells, M. (1989), *The Informational City,* London: Blackwell.

Castells, M. (1997), *The Power of Identity. Vol. II of The Information Age: Economy, Society and Culture,* Oxford: Blackwell.

Castro, J. (1994), *Tres años de soledad,* Bogotá: Opticas Gráficas Ltda.

Castro, J. and Tuirán, R. (2000), 'Las remesas de los trabajadores emigrantes a Estados Unidos', *Comercio Exterior,* **50,** 318–333.

Catterall, B. (2003), 'Is it all coming together? Further thoughts on urban studies and the present crisis', *CITY,* **7** (3), 423–432.

Chamberlain, H. (1998), 'Civil Society with Chinese Characteristics', *The China Journal,* **39,** January, 69–82.

Chambers, S. and Kymlicka, W. (eds) (2002), *Alternative Conceptions of Civil Society,* Princeton: Princeton University Press.

Chandler, T. (1987), *Four Thousand Years of Urban Growth: A Historical Census,* Lewiston/Queenston, NY: St. David's University Press.

Chang, Y.-S., (2003), 'Mutual Help and Democracy in Korea', in Bell, D.A. and Chaibong, H. (eds), *Confucianism for the Modern World,* New York: Cambridge University Press.

Chant, S. (1999), 'Population, migration, employment and gender', in Gwynne, R. and Kay, C. (eds), *Latin America transformed: globalization and modernity,* London: Arnold, pp. 226–269.

Chase-Dunn, C. and Weeks, J., 'Measuring The Suburbanization of world cities with Remote Sensing Data', paper presented at the Specialist Meeting on Globalization in the World-System: Mapping Change over Time, University of California Riverside, February 7–8, 2004, http://www.irows.ucr.edu/conferences/globgis/papers/Chase-Dunn_Weeks.htm, retrieved on February 9, 2004.

Cheek, T. (1998), 'From Market to Democracy in China – Gaps in the Civil Society Model', in Lindau, J.D. and Cheek, T. (eds), *Market Economics and Political Change: Comparing China and Mexico,* Lanham: Rowman & Littlefield Publishers.

Chesnais, J.-C. (1992), *The Demographic Transition: Stages, Patterns and Economic Implications,* New York: Oxford University Press.

Choe, S.-C. (1996), 'The evolving urban system in North-East Asia', in Lo, F.-C. and Yeung, Y.-M. (eds), *Globalization and the World of Large Cities,* Tokyo: UNU Press, pp. 498–519.

Chudacoff, H. P. (1981), *The Evolution of American Urban Society,* Englewood Cliffs: Prentice-Hall Inc.

CIA (2000), *The World Factbook 2000,* Washington D.C.

CIFAL (2004), *The Seven Principles of the Sustainable City.* Geneva.

Cohen, J.E. (1995), *How many people can the Earth support?,* New York: W.W. Norton & Company.

Cohen, J.L. and Arato, A. (1994), *Civil Society and Political Theory.* Cambridge, MA: MIT Press.

Cohen, M.A. (1996), 'The hypothesis of urban convergence: Are cities in the North and South becoming more alike in an age of globalization?', in Cohen, M.A., Ruble, B.A., Tulchin, J.S. and Garland, A.M. (eds), *Urban Future: Global Pressures and Local Forces,* Washington DC: Woodrow Wilson Center Press, pp. 25–38.

Costa, L. (1980) [1930], 'Razões da nova arquitetura', *Arte em Revista,* **4,** 15–23.

Cowherd, R. (2002), 'Cultural Construction of Jakarta: Design, Planning and Development in Jabotabek, 1980–1997', unpublished doctoral dissertation, Massachusetts Institute of Technology, Australia National University, Department of Architecture.

Crafts, N. (2000), 'Globalization and growth in the twentieth century', in IMF (ed.), *World Economic Outlook, Supporting Studies,* Washington, DC: International Monetary Fund.

Dahrendorf, R. (1994), *Life chances,* Chicago: University of Chicago Press.

Daly, H.E. (1994), 'Sustainable Growth: An impossibility theorem', in Daly, H.E. and Townsend, K.N. (eds), *Valuing the Earth: Economics, Ecology, Ethics,* Cambridge: MIT Press.

Dang, S. (2002), 'Creating Cosmopolis: the end of mainstream', unpublished Masters Thesis, School of Community and Regional Planning, University of British Columbia.

Davis, M. (1990), *City of Quartz: Excavating the Future in Los Angeles,* London: Verso.

De Boeck, F. (2002), 'Kinshasa: tales of the "invisible city" and the second world', in Enwezor, O., Basualdo, C. *et al.* (eds), *Under Siege: Four African Cities – Freetown, Johannesburg, Kinshasa, Lagos,* Documenta 11, Platform 4, Ostfildern-Ruit: Hatje Cantz Publishers, pp. 243–286.

De la Rocha, M. (1994), *The resources of poverty: Women and survival in a Mexican city,* Oxford, UK/Cambridge, USA: Blackwell.

de Mattos, C. (1999), 'Santiago de Chile, globalización y expansión metropolitana: lo que existía sigue existiendo', *Revista EURE,* **25** (76), 29–56.

de Roo, G. (2004), 'Coping with the growing complexity of our physical environment: the search for new planning tools in the Netherlands', in Sorenson, A., Marcotullio, P. J. and Grant, J. (eds), *Towards Sustainable Cities, East Asia, North American and European Perspectives on Managing Urban Regions,* Aldershot, UK: Ashgate Publishers, pp. 161–175.

De Soto, H. (2000), *The Mystery of Capital: Why Capitalism Triumphs in the West and Fails Everywhere Else,* New York: Basic Books.

Dean, W. (1969), *The Industrialization of São Paulo 1880–1945,* Austin: University of Texas Press.

Demeny, P. (1990), 'Population', in Turner, B.L. *et al.* (eds), *The Earth as Transformed by Human Action, Global and Regional Changes in the Biosphere over the last 300 Years,* Cambridge: Cambrige University Press, pp. 41–54.

Devas, N. (2004), 'Urban government: capacity, resources and responsiveness', in Devas, N. *et al.* (eds), *Urban Governance, Voice and Poverty in the Developing World,* London: Earthscan.

Devas, N. and Rakodi, C. (eds) (1993), *Managing Fast Growing Cities: New Approaches to Urban Planning and Management in the Developing World*, Harlow UK: Longman.

Dick, H.W. and Rimmer, P.J. (1998), 'Beyond the Third World City: the new urban geography of South-east Asia', *Urban Studies*, **35**, 2303–2321.

Dicken, P. (1998), *Global Shift: Transforming the World Economy*, London: Paul Chapman.

Ding, X.L. (1994), 'Institutional Amphibiousness and the Transition from Communism: The Case of China', *British Journal of Political Science*, **24** (1), 293–318.

Diniz, C.C. (1994), 'Polygonized development in Brazil: neither decentralization nor continued polarization', *International Journal of Urban and Regional Research*, **18**, 293–314.

Douglass, M. (1995), *Urban Environmental Management at the Grass Roots. Toward a Theory of Community Activation*, Honolulu: East-West Center Working Papers No. 42.

Douglass, M. (1998), 'A Regional Network Strategy for Reciprocal Rural-Urban Linkages', *Third World Planning Review*, **20** (1), 1–33.

Douglass, M. and Friedmann, J. (eds) (1998), *Cities for Citizens: planning and the rise of civil society in a global age*, Chichester UK: John Wiley & Sons.

Doxiadis, C.A. and Papioannou, J.G. (1974), *Ecumenopolis, the Inevitable City of the Future*, New York: W.W. Norton & Company, Inc.

Drost, A. (1996), 'Developing Sustainable Tourism for World Heritage Sites', *Annals of Tourism Research*, **23** (2), 479–492.

Drucker, P. (1986), 'The changed world economy', *Foreign Affairs*, **64**, 768–791.

Dugger, C. (2004), 'To Help Poor be Pupils, Not Wage Earners, Brazil Pays Parents', *New York Times*, January 3, p. A1.

Dunn, K., Hanna, B. and Thompson, S. (2001a), 'The Local Politics of Difference: An examination of inter-communal relations policy in Australian local government', *Environment and Planning A*, **33**, 1577–1595.

Dunn, K., Thompson, S., Hanna, B. and Burnley, I. (2001b), 'The Institution of Multiculturalism within Local Government in Australia', *Urban Studies*, **38** (13), 2477–2494.

Dunn, S. (ed). (1994), *Managing Divided Cities*, Staffs UK: Keele University Press.

Edgington, D. and Hutton, T. (2002), *Multiculturalism and Local Government in Greater Vancouver*, RIIM Working Paper Series, No. 02-06, University of British Columbia.

Edwards, M. and Hulme, D. (eds) (1996), *Beyond the Magic Bullet: NGO Performance and Accountability in the Post-Cold War World*, West Hartford, CN: Kumarian Press.

Edwards, S. (1995), *Crisis and reform in Latin America*, Oxford: Oxford University Press.

Ehrenreich, B. and Hochschild, A. (eds) (2003), *Global Women*, New York: Metropolitan Books.

Ehrlich, P.R. and Ehrlich, A.H. (1990), *The Population Explosion*, New York: Simon and Schuster.

Ehrlich, P.R. and Holdren, J.P. (1971), 'Impact of population growth', *Science*, **171**: 1212–1217.

Ehrlich, P.R. (1968), *The population bomb*, New York: Ballantine Books.

Enwezor, O., Basualdo, C., *et al.* (eds) (2002*), Under Siege: Four African Cities – Freetown, Johannesburg, Kinshasa, Lagos*, Documenta 11, Platform 4, Ostfildern-Ruit: Hatje Cantz Publishers.

Escobar, A. and González de la Rocha, M. (1995), 'Crisis, restructuring and urban poverty in Mexico', *Environment and Urbanization*, **7** (1), 57–76.

Ethnic Affairs Commission of New South Wales (1993), *New South Wales Charter of Principles for a Culturally Diverse Society*, Sydney: Ethic Affairs Commission of New South Wales.

Fainstein, S., Gordon, I. and Harloe, M. (1993*), Divided City: Economic Restructuring and Social Change in London and New York*, New York: Blackwell.

Fay, M. and Opal, C. (2000), *Urbanization without Growth: A Not-so-Uncommon Phenomenon*, Washington DC: World Bank Policy Research WP 2412.

Feiner, J. and Salmerón, D. (2000), *Greater Kunming Area – Regional Development Scenarios,* Zurich: Division for Landscape and Environmental Planning, ORL Institute, Swiss Federal Institute of Technology Zurich.

Feiner, J. and Salmerón, D. (2000), *Risks and Chances of Spatial Planning in the People's Republic of China, exemplified by the Greater Kunming Area, Yunnan Province,* Zurich: Division for Landscape and Environmental Planning, ORL Institute, Swiss Federal Institute of Technology Zurich.

Feiner, J., Schmid, W., Salmerón, D. and Eggenberger, M. (2000), 'Towards a more Sustainable Settlement and Transportation Pattern of the Greater Kunming Area', *DISP* **143**.

Fenster, T. (1999a), 'On particularism and universalism in modernist planning: mapping the boundaries of social change', *Plurimondi*, **2,** 147–168.

Fenster, T. (ed.) (1999b), *Gender, Planning, and Human Rights,* London: Routledge.

Fernández-Arias, E. and Hausmann, R. (2000), *What's wrong with international financial markets?,* Inter American Development Bank, Research Department Working Paper No. 429.

Firman, T. (2004), 'New town development in Jakarta Metropolitan Region: a perspective of spatial segregation', *Habitat International*, **28**, 349–368.

Fix, M. (2001), *Parceiros da Exclusão*, São Paulo: Boitempo Editorial.

Foster, K.W. (2002), 'Embedded within State Agencies: Business Associations in Yantai', *The China Journal*, **47,** January, 41–65.

Freire, P. (1972), *Pedagogy of the Oppressed*. London: Sheed & Ward.

Frey, H. (1999), *Designing the City: Towards a More Sustainable Urban Form.* London: Spon Press.

Friedmann, J. (1986), 'The world city hypothesis', *Development and Change*, **17,** 69–83.

Friedmann, J. (1993), *Empowerment: The Politics of Alternative Development.* Malden, MA: Blackwell.

Friedmann, J. and Lehrer U. (1997), 'Urban Policy Responses to Foreign In-Migration. The Case of Frankfurt am Main', *Journal of the American Planning Association*, **63** (1), Winter, 61–78.

Friedmann, J. and Wolff, G. (1982), 'World City Formation: an agenda for research and action', *International Journal of Urban and Regional Research*, **6**, 309–343.

Frúgoli Jr., H. (2000), *Centralidade em São Paulo. Trajetórias, Conflitos e Negociações na Metrópole,* São Paulo: Cortez/Edusp/Fapesp.

GAWC, *Globalization and World Cities – Study Group and Network.* Available at http://www.lboro.ac.uk/gawc/.

Giampietro, M., Bukkens, S.G.F. and Pimentel, D. (1992), 'Limits to population size: Three scenarios of energy interaction between human society and ecosystems', *Population and Environment*, **14**, 109–131.

Gilbert, A. (1998), 'World cities and the urban future: The view from Latin America', in Fu-Chen, L. and Yue-Man, Y. (eds), *Globalization and the world of large cities*, Tokyo, New York, Paris: UNU Press, pp. 174–202.

Gilbert, A. and Koser, K. (2002), *The dissemination to potential asylum seekers of information about UK immigration and asylum policy and practice,* Interim report to the Home Office, mimeo.

Gilbert, A.G. (1992), 'Third World cities: housing, infrastructure and servicing', *Urban Studies*, **29**, 435–460.

Gilbert, A.G. (1993), 'Third World cities: the changing national settlement system', *Urban Studies*, **30**, 721–740.

Gilbert, A.G. (1998), *The Latin American city*, London: Latin America Bureau and Monthly Review Press (revised and expanded edition).

Gilbert, A.G. (2001), *Housing in Latin America,* Inter-American Development Bank, INDES-European Union Joint Program Working Paper.

Gilbert, A.G. (ed.) (1996), *The mega-city in Latin America*, Tokyo: United Nations University Press.

Gilbert, A.G. and Dávila, J. (2002), 'Bogotá: progress in a hostile environment', in Myers, D.J., Dietz, D.A. and Orsini J.-F.L. (eds), *Capital City politics in Latin America: Democratization and Empowerment*, Boulder, CO: Lynne Rienner, pp. 29–64.

Ginkel, H.J.A. van (1979), 'Suburbanisatie en Recente Woonmilieus', *Utrechtse Geografische Studies* **15, 16**.

Ginkel, H.J.A. van (1986), 'Population Growth and Settlement Pattern in the Green Heart of the Randstad', in Enyedi, G. and Veldman, J. (eds), *Rural Development Issues in Industrializes Countries*, Pecs: Center for Regional Studies, Hungarian Academy of Sciences.

Ginsburg, N., Koppel, B. and McGee, T.G. (eds) (1991), *The Extended Metropolis: Settlement Transition in Asia,* Honolulu: University of Hawaii Press.

Glaeser, E. (1998), 'Are Cities Dying?', *The Journal of Economic Perspectives*, **12** (2), Spring, 139–160.

Gold, T.B. (1990), 'The Resurgence of Civil Society in China,' *Journal of Democracy*, **1** (1), Winter, 18–31.

Gottman, J. (1961), *Megalopolis: The Urbanized Northeastern Seaboard of the United States,* New York: The Twentieth Century Fund, Krauss International Publications.

Graham, S. and Marvin, S. (1996*), Telecommunications and the City: electronic spaces, urban places*, London: Routledge.

Gramsci, A. (1971), *Selections from the Prison Notebooks,* New York: International Publishers.

Grassby, A. (1973), *A Multicultural Society for the Future,* Canberra: Department of Immigration/Australian Government Publishing Service.

Gret, M. and Sintomer, Y. (2004), *The Porto Alegre experiment: learning lessons for better democracy,* London: Zed Books.

Gu, X. (1993/4), 'A Civil Society and Public Sphere in Post-Mao China? An Overview of Western Publications', *China Information,* **8** (3), Winter, 38–52.

Guinness, P. (1986), *Harmony and hierarchy in a Javanese kampung,* Singapore, Oxford: Oxford University Press.

Guldin, G.E. (2001), *What's a peasant to do?: village becoming town in southern China,* Boulder, CO: Westview Press.

Gutierrez, G. (1973), *A Theology of Liberation: The Spiritual Journey of a People,* Maryknoll, NY: Orbis Books.

Habermas, J. (1989), *The Structural Transformation of the Public Sphere: An Inquiry into a Category of Bourgeois Society,* translated Burger, T. with the assistance of Lawrence, F., Cambridge: Polity Press. (Original German edition 1962).

Hack, G. (2000), 'Infrastructure and regional form', in Simmonds, R. and Hack, G. (eds), *Global City Regions, Their Emerging Forms,* London: Spon Press, pp. 183–192.

Hage, G. (1998), *White Nation: fantasies of white supremacy in a multicultural society,* Sydney: Pluto Press.

Hall, D.L. and Ames, R.T. (2003), 'A Pragmatist Understanding of Confucian Democracy', in Bell, D.A. and Chaibong, H. (eds), *Confucianism for the Modern World,* New York: Cambridge University Press, pp. 124–160.

Hall, D.R. (1998), 'Tourism Development and Sustainability Issues in Central and Southeastern Europe', *Tourism Management,* **19** (5), October, 423–431.

Hall, P. and Pfeiffer, U. (2000), *Urban Future 21: A Global Agenda for Twenty-First Century Cities,* New York: Spon Press.

Hall, S. (1991), 'The Local and the Global: Globalization and Ethnicity', in King, A.D. (ed.), *Culture, Globalization and the World-System: Contemporary Conditions for the Representation of Identity. Current Debates in Art History,* Binghamton: State University of New York, 3rd Department of Art and Art History.

Hardoy, J.E., Mitlin, D. and Satterthwaite, D. (2001), *Environmental Problems in an Urbanization World,* London: Earthscan.

Hare, D. and West, L.A. (1999), 'Spatial Patterns in China's Rural Industrial Growth and Prospects for the Alleviation of Regional Income Inequality', *Journal of Comparative Economics,* **27**, 475–497.

Harvey, D. (1989), *The Condition of Postmodernity: An Enquiry into the Origins of Cultural Change,* Oxford: Blackwell.

Hatch, W. and Yamamura, K. (1996), *Asia in Japan's Embrace, Building a Regional Production Alliance,* Hong Kong: Cambridge University Press.

Haughton, G. and Hunter, C. (1994), *Sustainable Cities,* London: Regional Studies Association.

Headrick, D. R. (1990), 'Technological change', in Turner, B. L. (eds.), *The Earth as Transformed by Human Action, Global and Regional Changes in the Biosphere over the last 300 Years,* Cambridge: Cambridge University Press, pp. 55–67.

Heinrich, C. (2001), *ICT Infrastructure and Externalities Affecting Spatial Structures*, Working Paper of the Globalization and World Cities Study Group and Network, http://www.lboro.ac.uk/gawc/rb/rb67.html, retrieved on October 4, 2003.

Held, D., McGrew, A., Goldblatt, D. and Perraton, J. (1999), *Global Transformations, Politics, Economics and Culture,* Cambridge: Polity Press.

Hiebert, D. (2003), *Are Immigrants Welcome? Introducing the Vancouver Community Studies Survey,* RIIM Working Paper, No. 03-06.

Higgins, B.H. and Savoie, D.J. (1997), *Regional Development Theories and Their Application,* New Brunswick, N.J. and London: Transaction.

Hirabe, N. (2003), *Multicultural Planning,* Class Research Paper, UBC, Fall, PLAN 548e.

Ho, S.P.S. and Lin, G.C.S. (2003), 'Emerging land markets in rural and urban China: policies and practices', *The China Quarterly,* **175,** 681–707.

Holston, J. (1989), *The Modernist City: An Anthropological Critique of Brasília,* Chicago: University of Chicago Press.

Holston, J. (1991a), 'Autoconstruction in working-class Brazil', *Cultural Anthropology,* **6** (4), 447–465.

Holston, J. (1991b), 'The misrule of law: land and usurpation in Brazil', *Comparative Studies in Society and History,* **33** (4), 695–725.

Holston, J. (1999), *Cities and Citizenship,* Durham: Duke University Press.

Holston, J. (2001), 'The spirit of Brasília: modernity as experiment and risk', in Sullivan, E.J. (ed.), *Brazil Body & Soul,* New York: The Solomon R. Guggenheim Museum, pp. 540–557.

Hondagneu-Sotelo, P. (1994), *Gendered Transitions,* Berkeley: University of California Press.

Honjo, M. (1998), 'The growth of Tokyo as a world city', in Fu-Chen, L. and Yue-Man, Y. (eds), *Globalization and the world of large cities,* Tokyo, New York, Paris: United Nations University Press, pp. 109–131.

Hösle, V. (1999), 'Gerechtigkeit zwischen den Generationen', in Gräfin Dönhoff, M. and Sommer, T. (eds), *Was steht uns bevor? Mutmaßungen über das 21. Jahrhundert, Aus Anlaß des 80. Geburtstages von Helmut Schmidt,* Berlin: Siedler, pp. 189–200.

Hough, M. (1989), *City Form and Natural Process,* London: Croom Helm.

Huang, P. (1993), '"Public Sphere"/"Civil Society" in China?', *Modern China,* **19,** (2), April, 217–240.

IBGE Instituto Brasileiro de Geografia e Estatística (1959), *Censo Experimental de Brasília,* Rio de Janeiro: IBGE.

Iglesias, E.V. (1992), *Reflections on economic development: toward a new Latin American consensus,* Inter-American Development Bank.

ILO International Labor Office (1995), *World employment: an ILO report: 1995,* Geneva.

Institute for Sustainable Communities, *Definitions of a sustainable community,* http://www.iscvt.org/ FAQscdef.htm, retrieved on February 22, 2004.

Inter American Development Bank (1999), *Facing up to inequality, Economic and Social Progress in Latin America,* Inter-American Development Bank Annual report 1998-9, Washington D.C.

Jackson, S. (1998), 'Geographies of Coexistence: native title, cultural difference and the decolonization of planning in north Australia', unpublished PhD dissertation, School of Earth Sciences, Macquarie University, Sydney.

Jacobs, J. (1969), *The Economy of Cities*, New York: Random House.

Jacobs, J. (1984), *Cities and the Wealth of Nations*, New York: Random House.

Jacobs, J. (1996), *Edge of Empire. Postcolonialism and the City*, London: Routledge.

Janelle, D.G. (1968), 'Central place development in a time-space framework', *Professional Geographer,* **20**, 5–10.

Janelle, D.G. (1969), 'Spatial reorganization: a model and concept', *Annals of the Association of American Geographers,* **59**, 348–364.

Johnston, R.J., Taylor, P.J. and Watts, M.J. (1995), *Geographies of Global Change: Remapping the World in the Late Twentieth Century,* Oxford: Blackwell.

Jokisch, B. and Pribilsky, J. (2002), 'The panic to leave: economic crisis and the "new emigration" from Ecuador', *International Migration,* **40**, 74–99.

Jonas, H. (1985), *The Imperative of Responsibility – In Search of an Ethics for the Technological Age*, Chicago: University of Chicago Press.

Jones, G.W., Douglass, R.M., Caldwell, J.C., D'Souza, R.M. (eds) (1997), *The Continuing Demographic Transition*, New York: Oxford University Press.

Kant, I. (1788, Reprint 2003), *Kritik der reinen Vernunft. Kritik der praktischen Vernunft. Kritik der Urteilskraft*, Wiesbaden: Fourierverlag.

Kates, R.W., Billie, L., Turner, I. and Clark, W. (1990), 'The great transformation', in Years ,W. *et al.* (eds), *The Earth as Transformed by Human Action, Global and Regional Changes in the Biosphere over the last 300,* Cambridge: Cambridge University Press, pp. 1–17.

Keane, J. (1998), *Civil Society: Old Images, New Visions*. Oxford: Polity Press.

Keiner, M. (2002), 'Indicator based control of regional planning', *Australian Planner,* **39** (4), December, 205–210.

Keiner, M. (2004), 'Re-Emphasizing Sustainable Development – The concept of Evolutionability'. On living chances, equity and good heritage, *Environment, Development and Sustainability,* **6** (4), 379–392.

Keiner, M. and Schmid, W.A. (2003), 'Urbanisierungstendenzen in Entwicklungsländern. Probleme und Potenziale für nachhaltige Stadtentwicklung', *DISP* **155**, 49–56.

Keiner, M., Salmerón, D., Schmid, W.A. and Poduje, I. (2004), 'Urban Development in Southern Africa and Latin America', in Keiner, M., Zegras, C., Schmid, W.A. and Salmerón, D. (eds), *From Understanding to Action – Sustainable Urban Development in Medium-Sized Cities in Africa and Latin America*, Dordrecht: Springer, 1–24.

Kelly, P.F. (2000), *Landscapes of globalization: human geographies of economic change in the Philippines,* London and New York: Routledge.

Kenworthy, J., Laube, F., Newman, P. and Barter, P. (1997), 'Indicators of Transport Efficiency in 37 Global Cities', A report for the World Bank, Perth: Murdock University, Sustainable Transport Research Group.

King, A. (1976), *Colonial Urban Development: Culture, Social Power, and Environment,* London: Routledge and Kegan Paul.

King, A.D. (1990), *Urbanism, Colonialism, and the World Economy. Culture and Spatial Foundations of the World Urban System*, The International Library of Sociology, London and New York: Routledge.

King, A.D. (ed.) (1996), *Representing the City. Ethnicity, Capital and Culture in the 21st Century,* London: Macmillan.

King, R., Inkoom, D., *et al.* (2001), *Urban Governance in Kumasi: Poverty and Exclusion,* Urban Poverty and Governance Working Chapter No. 23, Birmingham: International Development Department, School of Public Policy, University of Birmingham.

Knox, P. and Agnew, J. (1998), *The Geography of the World Economy,* London: Arnold.

Knox, P.L. and Taylor, P.J. (eds) (1995), *World Cities in a World-System*, Cambridge, UK: Cambridge University Press.

Koh, D.W.H. (2000), 'Wards of Hanoi and state-society relations in the Socialist Republic of Vietnam', unpublished doctoral dissertation, Department of Political and Social Change, Australia National University, Canberra.

Kojima, K. (2000), 'The 'flying geese' model of Asian economic development: origin, theoretical extensions, and regional policy implications', *Journal of Asian Economies,* **11**, 375–401.

Kombe, W.J. and Kreibich, V. (2000), *Informal Land Management in Tanzania,* SPRING Research Series No. 29, University of Dortmund.

Kondratieff, N.D. (1979), 'The long waves of economic life', *Review II:* 519–562.

Korboe, D., Diaw, K., *et al.* (2000), *Urban Governance, Partnership and Poverty: Kumasi,* Urban Governance, Partnership and Poverty Working Chapter No. 10, Birmingham: International Development Department, School of Public Policy, University of Birmingham.

Kubitschek, J. (1975), *Por Que Construí Brasília*, Rio de Janeiro: Bloch Editores.

Lachaud, J.-P. (1994), *The African Labour Market,* Geneva: ILO.

Le Corbusier (Charles Edouard Jeanneret) (1957) [1941], *La Charte d'Athènes*, Paris: Editions de Minuit.

Leaf, M. (1991), 'Land Regulation and Housing Development in Jakarta, Indonesia: From the "Big Village" to the "Modern City"', unpublished doctoral dissertation, Department of City and Regional Planning, University of California, Berkeley.

Leaf, M. (1993), 'Land Rights for Residential Development in Jakarta, Indonesia: The Colonial Roots of Contemporary Urban Dualism', *International Journal of Urban and Regional Research,* **17** (4), 477–491.

Leaf, M. (1994), 'Legal Authority in an Extralegal Setting: The Case of Land Rights in Jakarta, Indonesia', *Journal of Planning Education and Research,* **14,** 12–18.

Leaf, M. (1999), 'Vietnam's Urban Edge: The Administration of Urban Development in Hanoi', *Third World Planning Review,* **21** (3), 297–315.

Leaf, M. (2002), 'A Tale of Two Villages: Globalization and Peri-Urban Change in China and Vietnam', *Cities,* **19** (1), 23–31.

Leaf, M. (2005a), 'The Bazaar and the Normal: Informalization and Tertiarization in Urban Asia', in Daniels, P.W., Ho, K.C. and Hutton, T.A (eds), *Service Industries and Asia Pacific Cities: New Development Trajectories,* London and New York: Routledge.

Leaf, M. (2005b), 'Modernity Confronts Tradition: the Professional Planner and Local Corporatism in the Rebuilding of China's Cities', in Sanyal, B. (ed.), *Comparative Planning Cultures,* London, New York: Routledge.

LeGates, R.T. and Stout, F. (eds) (1996), *The City Reader,* London and New York: Routledge.

Levine, D.H. and Mainwaring, S. (1989), 'Religion and Popular Protest in Latin America', in Eckstein, S. (ed.), *Power and Popular Protest: Latin American Social Movements,* Berkeley: University of California Press, pp. 203–240.

Lew, S.-C., Chang, M.-H. and Kim, T.-E. (2003), 'Affective Networks and Modernity: The Case of Korea', in Bell, D.A. and Chaibong, H. (eds), *Confucianism for the Modern World,* New York: Cambridge University Press, pp. 201–217.

Lin, G. C.-S. (1994), 'Changing theoretical perspectives on urbanization in Asian developing countries', *Third World Planning Review* **16,** 1–23.

Lins, P. (1997), *Cidade de Deus,* São Paulo: Companhia das Letras.

Liu, J., Daily, G.C., Ehrlich, P.R. and Luck, G.W. (2003), 'Effects of household dynamics on resource consumption and biodiversity', *Nature,* **421,** 530–533.

Lloyd-Sherlock, P. (1997), 'The Recent Appearance of Favelas in Sao Paulo City: An Old Problem in a New Setting', *Bulletin of Latin American Research,* **16** (3), 289–305.

Lo, F.-C. (1994), 'The impacts of current global adjustment and shifting techno-economic paradigm on the world city system', in Fuchs, R.J., Brennan, E., Chamie, J., Lo, F.-C. and Uitto, J.I. (eds), *Mega-City Growth and The Future,* Tokyo: United Nations University Press, pp. 103–130.

Lo F.-C. & Marcotullio P.J. (2000), 'Globalization and urban transformations in the Asia Pacific region: a review', *Urban Studies,* **37,** 77–111.

Lo, F.-C. and Marcotullio, P.J. (eds) (2001), *Globalization and the Sustainability of Cities in the Asia Pacific Region,* Tokyo: United Nations University Press.

Lo, F.-C. and Yeung, Y.-M. (eds) (1996), *Emerging World Cities in Pacific Asia,* Tokyo: United Nations University Press.

Lo, F.-C. and Yeung, Y.-M. (eds) (1998), *Globalization and the World of Large Cities,* Tokyo: United Nations University Press.

Logan, J.R. (2001), *The new latinos: who they are, where they are?,* Lewis Mumford Center for Comparative Urban and Regional Research, Albany: New York University.

Lomnitz, L. (1977), *Networks and Marginality: Life in a Mexican shantytown,* New York, San Francisco, London: Academic Press.

Lutz, W., Sanderson, W.C. and Scherbov, S. (eds) (2004), *The End of World Population Growth in the 21st Century, New Challenges for Human Capital Formation and Sustainable Development,* London: Earthscan.

Mabin, A. and Smit, D. (1997), 'Reconstructing South Africa's cities? The making of urban planning 1900–2000', *Planning Perspectives,* **12,** 193–223.

Maddison, A. (1991), *Dynamic Forces in Capitalist Development: A Long-Run Comparative View,* Oxford: Oxford University Press.

Maddison, A. (2001), *The World Economy, A Millennial Perspective,* Paris: OECD.

Mahtani, M. (2002), 'Interrogating the Hyphen-Nation: Canadian Multicultural Policy and "Mixed Race" Identities', *Social Identities,* **8** (1), 67–90.

Malthus, T. (1798), *An Essay on the Principle of Population*, London, Online book. http://www.econlib.org/library/Malthus/malPop.html. Liberty Fund, Inc. Indianapolis: The library of Economics and Liberty.

Marcotullio, P.J. (2003), 'Globalization, urban form and environmental conditions in Asia Pacific cities', *Urban Studies, 40,* 219–248.

Marcotullio, P.J. (2004), *Time-space telescoping and Asian urbanization,* Yokohama: United Nations University Institute of Advanced Studies.

Marcotullio, P.J. and Lee Y.-S.F. (2003), 'Urban environmental transitions and urban transportation systems: a comparison of North American and Asian experiences', *International Development Planning Review, 25,* 325–354.

Marcotullio, P.J., Rothenberg, S. and Nakahari, M. (2003), 'Globalization and urban environmental transitions: comparison of New York's and Tokyo's experiences', *Annals of Regional Science, 37,* 369–390.

Marcotullio, P.J., Williams, E.W. and Marshall, J.D. (2004), *Faster, sooner, and more simultaneously: how recent transportation CO_2 emission trends in developing countries differ from historic trends in the United States of America,* Yokohama: United Nations University Institute of Advanced Studies.

Marcuse, P. (1998), 'Sustainability is not enough', *Environment and Urbanization,* **10** (2), 103–111.

Martin, J. and Warner, S.B. (2000), 'Local initiative and Metropolitan Repetition: Chicago 1972–1990', in Fishman, R. (ed.), *The American Planning Tradition,* Washington, DC: Woodrow Wilson Center Press.

Marton, A.M. (2000), *China's spatial economic development: restless landscapes in the lower Yangzi Delta,* London and New York: Routledge.

McGee, T.G. (1991), 'The emergence of desakota regions in Asia: expanding a hypothesis', in Ginsberg, N., Koppel, B. and McGee, T.G. (eds), *The Extended Metropolis: Settlement Transition in Asia,* Honolulu: University of Hawaii Press, 3–25.

McGee, T.G. and Robinson, I.M. (eds) (1995), *The Mega-Urban Regions of Southeast Asia,* Vancouver: University of British Columbia Press.

McGranahan, G. and Satterthwaite, D. (2003), 'Urban Centers: An Assessment of Sustainability', *Annual Review of Environment and Resources,* **28,** 243–274.

McGranahan, G., Jacobi, P., Songsore, J., Surjadi, C. and Kjellen, M. (2001), *The Citizens at Risk, From Urban Sanitation to Sustainable Cities,* London: Earthscan.

Meadows D.H, Meadows, D.L., Randers, J. and Behrens, W.W. (1972), *The limits to growth*, New York: Universe books.

Melosi, M. (1999), 'Refuse pollution and municipal reform: the waste problem in America, 1880–1917', in Roberts, G.K. (ed.), *The American Cities and Technology Reader,* London: Routledge and the Open University, pp. 163–172.

Melosi, M. V. (2000), *The Sanitary City: Urban Infrastructure in America from Colonial Times to the Present,* Baltimore: Johns Hopkins Press.

Meyer, R.M.P. (1991), 'Metrópole e Urbanismo: São Paulo Anos 50', Ph.D. Dissertation, Universidade de São Paulo, Faculdade de Arquitetura e Urbanismo.

Meyer, W.B. (1996), *Human Impact on the Earth,* Cambridge: Cambridge University Press.

Millennium Ecosystem Assessment (2003), *Ecosystems and Human Well-being,* Washington, DC: Island Press.

Miller, G. (2001), 'The Development of Indicators for Sustainable Tourism: Results of a Delphi Survey of Tourism Researchers', *Tourism Management,* **22** (4), August, 351–362.

Milroy, B. and Wallace, M. (2001), 'Ethno-racial diversity and planning practices in the Greater Toronto Area', *Plan Canada* **41** (3), 31–33.

Ministerie van Volkshuisvesting, Ruimtelijke Ordening en Milieubeheer – Rijks Planologische Dienst (2001), *Ruimte maken, Ruimte delen. Vijfde Nota over de Ruimtelijke Ordening 2000/2020*, januari 2001, Den Haag.

Ministry of Justice (1959), *Brasília: Medidas Legislativas Sugeridas à Comissão Mista pelo Ministro da Justiça e Negócios Interiores*, Rio de Janeiro: Departamento de Imprensa Nacional.

Miura F. (2000), 'Characteristics of Overseas trends in measures for Aging Society with major focus on trends in six foreign countries', in Policy Office of the Aging of Society, Director-Secretariat of the Management and Coordination Agency, Tokyo (ed), *The Status of Aging in the Population and Measures for an Aged Society in Various Countries*, pp. 3–19.

Molina, L.T. and Molina, M.J. (eds) (2002), *Air Quality in the Mexico Megacity, An integrated assessment,* Alliance for Global Sustainability Book series Science and Technology: Tools for Sustainable Development, Volume 2, Dordrecht, Boston, London: Kluwer.

Montgomery, M.R., Stren, R. *et al.* (2004), *Cities Transformed: Demographic Change and Its Implications in the Developing World*, National Academic Press.

Morse, R.M. (1970), *Formação Histórica de São Paulo*, São Paulo: Difel.

Moser, C., Norton, A., Conway, T., Ferguson, C. and Vizard, P. (2001), *To Claim our Rights: Livelihood Security, Human Rights and Sustainable Development*, Overseas Development Institute.

Moser, C.O.N. (1984), 'The Informal Sector Reworked: Viability and Vulnerability in Urban Development', *Regional Development Dialogue,* **5** (3), 135–185.

Muldavin J. (1998), 'The limits of market triumphalism in rural China', *Geoforum,* **28,** (3-4), 289–312.

Munck, R. (2003), *Contemporary Latin America*, Hampshire: Palgrave Macmillan.

Myers, D.J., Dietz, D.A. and Orsini J.-F. L. (eds) (2002*), Capital City Politics in Latin America: Democratization and Empowerment*, Boulder, CO: Lynne Rienner.

NASA (2001), http://www.gsfc.nasa.gov/gsfc/earth/pictures/citylights/flat_earth _nightm.jpg, retrieved on February 28, 2004.

Nelson, J. (1969), *Migrants, Urban Poverty and Instability in Developing Nations*, Center for International Affairs, Occasional Chapters in International Affairs, No. 22, Cambridge, MA: Harvard University.

Ness, G.D. (with Low, M.) (2000), *Five Cities: Modelling Asian Urban Population-Environment Dynamics*, New York: Oxford University Press.

New York Times (2004), *Brazil Adopts Strict Gun Controls to Try to Curb Murders*, January 21, p. A3.

Nlandu, T. (2002), 'Kinshasa: beyond chaos', in Enwezor, O., Basualdo, C., Baueret U.M. *et al.* (eds), *Under Siege: Four African Cities – Freetown, Johannesburg, Kinshasa, Lagos*. Documenta 11, Platform 4, Ostfildern-Ruit: Hatje Cantz Publishers, pp. 185–200.

Noyelle, T. and Dutka, A.B. (1988), *International Trade in Business Services: Accounting, Advertising, Law and Management Consulting*, Cambridge, MA: Ballinger Publishing.

O'Meara, M. (1999), 'Reinventing cities for people and the planet', *Worldwatch Paper* **147**, Worldwatch Institute.

One World Action (2002), *From Consultation to Influence: Citizen Voices, Responsiveness and Accountability in Service Delivery*, London: One World Action.

Paoli, M.C. (1999), 'Apresentação e introdução', in Paoli, M. C. and de Oliveira, F. (eds), *Os Sentidos da Democracia – Políticas do Dissenso e Hegemonia Global*, São Paulo: Nedic, Fapesp and Editora Vozes, pp. 7–23.

Papademetriou, D. (2002), *Reflections on International Migration and its Future*, Kingston, Ontario: Queens University, School of Policy Studies, The J. Douglas Gibson Lecture.

Parnreiter, C, Komlosy, A., Stacher, I. and Zimmermann, S. (1997), *Ungeregelt und unterbezahlt. Der informelle Sektor in der Weltwirtschaft*. Frankfurt/Main: Brandes & Apsel/Südwind.

Payne, J. and Majale, M. (2004), *The Urban Housing Manual, Making Regulatory Frameworks Work for the Poor*, London: Earthscan.

Pei, M. (1998), 'Chinese Civic Association: An Empirical Analysis', *Modern China*, **24** (3), 285–318.

Pelling, M. (2003), *The Vulnerability of Cities: Natural Disaster and Social Resilience*, London: Earthscan.

Perlman, J. (1976), *The Myth of Marginality: Urban Politics and Poverty in Rio de Janeiro*, Berkeley: University of California Press.

Perlman, J. (1993), *Mega-cities: Global urbanization and innovation*, The Mega-Cities Project Publication MCP-013, Hartford.

Perlman, J. (2004), 'Marginality: From Myth to Reality in the Favelas of Rio de Janeiro: 1969–2002', in Roy, A. and Alsayyad, N. (eds), *Urban Informality: Transnational Perspectives from the Middle East, Latin America, and South Asia*, New York: Lexington.

Perlman, J., Hopkins, E. and Jonsson, Å. (1998), *Urban solutions at the poverty/environment intersection*, The Mega cities Project Publication MCP-018, Hartford.

Perlman, J.E. (1980), *The Myth of Marginality*. University of California Press.

Pero, V., Cardoso, A. and Elias, P. (2003), *Urban Regeneration and spatial discrimination: the case of Rio's favelas*, Anais do XXXI Encontro Nacional de Economia (Proceedings of the 31th Brazilian Economics Meeting. Brazilian Association of Graduate Programs in Economics ANPEC).

Picatto, P. (2003), 'Understanding crime in twentieth-century', *Enfoque*, Fall, 1–6.

Piermay, J.-L. (1997), 'Kinshasa: a reprieved mega-city?' in Rakodi, C. (ed.), *The Urban Challenge in Africa: Growth and Management of Its Large Cities*, Tokyo: United Nations University Press, pp. 223–251.

Pimentel, D. (1999), 'How many people can the Earth support?' *Population Press*, **5** (3), March/April, The Population Coalition.

Portes, A. (1989), 'Latin American urbanization during the years of the crisis', *Latin American Research Review*, **25**, 7–44.

Portes, A., Castells, M., and Benton, L.A. (eds) (1989), *The Informal Economy: Studies in Advanced and Less Developed Countries*, Baltimore: Johns Hopkins University Press.

Potts, D. (1997), 'Urban lives: adopting new strategies and adapting rural links', in Rakodi, C. (ed.), *The Urban Challenge in Africa: Growth and Management of Its Large Cities*, Tokyo: United Nations University Press, pp. 447–496.

PRC National Bureau of Statistics (1999), *China Statistical Yearbook 1999*, Beijing: China Statistical Publishing House.

Prud'homme, R. (1996), 'Managing Megacities', *Le courrier du CNRS*, **82**, 174–176.

Putnam, R. (1994), *Making Democracy Work: Civic Traditions in Modern Italy*, Princeton: Princeton University Press.

Rabinow, P. (1989), *French Modern. Norms and Forms of the Social Environment*, Cambridge, MA: MIT Press.

Rakodi, C. (1997), 'Residential property markets in African cities', in Rakodi, C. (ed.), *The Urban Challenge in Africa*, Tokyo: United Nations University Press, pp. 371–410.

Rakodi, C. (2004), 'Urban politics: exclusion or empowerment?' in Devas, N. *et al.* (eds), *Urban Governance, Voice and Poverty in the Developing World*, London: Earthscan.

Rakodi, C. (ed.) (1997), *The Urban Challenge in Africa*, Tokyo, New York, Paris: UNU Press.

Rakodi, C., Gatabaki-Kamau, R. and Devas, N. (2000), 'Poverty and political conflict in Mombasa', *Environment and Urbanization*, **11** (2), 153–170.

Riddell, B. (1997), 'Structural adjustment programmes and the city in tropical Africa', *Urban Studies*, **34** (8), 1297–1307.

Rigg, J. (2003), *Southeast Asia: The Human Landscape of Modernization and Development*, second edition, London: Routledge.

Roberts, B. (1994), 'Informal economy and family strategies', *International Journal of Urban and Regional Research*, **18**, 6–23.

Rocha, J. (2003), ''Cutting the Wire': The Landless Movement in Brazil', *Current History*, **102** (661), February, 86–90.

Rogerson, C. (1997), 'Globalization or informalization? African urban economies in the 1990s', in Rakodi, C. (ed.), *The Urban Challenge in Africa: Growth and Management of Its Large Centers*. Tokyo: United Nations University Press, pp. 337–370.

Rolnik, R. (1997), *A Cidade e a Lei: Legislaçao, Política Urbana e Territórios na Cidade de São Paulo*, São Paulo: FAPESP/Studio Nobel.

Rotker, S. (ed.) (2002), *Citizens of fear: urban violence in Latin America*, New Brunswick, NJ: Rutgers University Press.

Rousseau, J.-J. (1762, Reprint 2003), *On the social contract*, Mineola, NY: Dover Publications.

Rozelle, S., Huang, J. and Zhang, L. (1997): 'Poverty, Population and Environmental Degradation in China', *Food Policy*, **22** (3), 229–251.

Sahn, D.E. and Stifel, D.C. (2003), 'Progress Toward the Millennium Development Goals in Africa', *World Development*, **31** (1), 23–52.

Sandercock, L. (1997), *Towards Cosmopolis: Planning for Multicultural Cities*, New York: John Wiley & Sons.

Sandercock, L. (2002), 'Practicing Utopia: Sustaining cities', *DISP,* **148**, 4–9.
Sandercock, L. (2003), *Cosmopolis 2: Mongrel Cities of the 21st Century*, London: Continuum Books.
Sandercock, L., Dickout, L. and Winkler, T. (2004), *The Quest for an Inclusive City: an exploration of the Sri Lankan Tamil experience of integration in Toronto and Vancouver*, Working paper 04-12, Vancouver Centre of Excellence, Research on Immigration and Integration in the Metropolis, May 2004.
Santos, M. (1996), 'São Paulo: A growth process full of contradictions', in Gilbert, A. (ed.), *The Mega-City in Latin America,* Tokyo, New York, Paris: UNU Press, pp. 224–240.
Sassen S. (1991), *The Global City*: New York, London, Tokyo, Princeton: Princeton University Press.
Sassen, S. (1988) *The Mobility of Labor and Capital. A study in international investment and capital flow*, Cambridge: Cambridge University Press.
Sassen, S. (1991), *The Global Cities*, Princeton: Princeton University Press.
Sassen, S. (1996), *Losing Control? Sovereignty in an Age of Globalization*, New York: Columbia University Press.
Sassen, S. (1998), *Globalization and its Discontents,* New York: New Press.
Sassen, S. (2000), 'The Demise of Pax Americana and the Emergence of Informalization as a Systemic Trend', in Tabak, F. and Crichlow, M.A. (eds), *Informalization: Process and Structure*, Baltimore: Johns Hopkins University Press, pp. 91–115.
Sassen, S. (2000), *Guests and Aliens*, New York: The New Press.
Sassen, S. (2000a), *Cities in a World Economy*, 2nd edn, Pine Forge: Sage Press.
Sassen, S. (2001), *The Global City: New York, London, Tokyo*, Princeton University Press.
Sassen, S. (2002), *Global Networks/Linked Cities*, New York and London: Routledge.
Sassen, S. (2003), 'The Repositioning of Citizenship: Emergent Subjects and Spaces for Politics', *Berkeley Journal of Sociology*, **46**, 4–26.
Satterthwaite, D. (1996), 'Towards healthy cities', *People and the Planet,* **5** (2).
Satterthwaite, D. (2003), 'The Millenium Development Goals and urban poverty reduction: great expectations and nonsense statistics', *Environment and Urbanization,* **15** (2), 181–190.
Schell, O. (2004), 'China's Hidden Democratic Legacy', *Foreign Affairs*, **83**, 116–124.
Seeborg, M.C., Jin, Z. and Zhu, Y. (2000), 'The New Rural-Urban Mobility in China: Causes and Implications', *Journal of Socio-Economics,* **29**, 39–56.
Short, J.R. and Kim, Y. (1999), *Globalization and the City*, Essex: Longman.
Silva, A.A. (1990), 'A luta pelos direitos urbanos: novas representações de cidade e cidadania', *Espaço e Debates* **30**, 29–41.
Simon, D. (1997), 'Urbanization, globalization and economic crisis in Africa', in Rakodi, C. (ed.), *The Urban Challenge in Africa: Growth and Management of its Large Cities,* Tokyo: United Nations University Press, pp. 74–109.
Simon, J. (1981), *The Ultimate Resource,* Princeton, NJ: Princeton University Press.
Simon, J. (1998), *The Ultimate Resource 2*, Princeton, NJ: Princeton University Press.

Simone, A. (2002), *Principles and Realities of Urban Governance in Africa*, Nairobi: UN-HABITAT.

Singer, P. (1984), 'Interpretação do Brasil: uma experiência histórica de desenvolvimento', in Boris F. (ed.), *História Geral da Civilização Brasileira, vol. 2: O Brasil Republicano, 4 Economia e Cultura (1930–1964)*, São Paulo: Difel, pp. 211–245.

Smith K.R. and Lee Y.-S.F. (1993), 'Urbanization and the environmental risk transition', in Kasarda, J.D. and Parnell, A.M. (eds), *Third World Cities: Problems, Policies, and Prospects*, Newbury Park: Sage Publications, pp. 161–179.

Smith, K. (1990), 'The risk transition', *International Environmental Affairs*, **2**, 227–251.

Solinger, D. (1999), *Contesting Citizenship in Urban China: Peasant Migrants, the State and the Logic of the Market*, Berkeley: University of California Press.

Somekh, N. (1992), 'Plano Diretor de São Paulo: uma aplicação das propostas de solo criado', in de Queiroz Ribeiro, L.C. and do Lago, L.C. (eds), *Acumulação Urbana e a Cidade*, Rio de Janeiro: IPPUR/UFRJ, pp. 255–260.

Somekh, N. and Campos, C.M. (eds) (2002), *A Cidade que Não Pode Parar: Planos Urbanísticos de São Paulo no Século XX*, São Paulo: Mack Pesquisa.

Southworth, F.A. (1995), 'Technical Review of Urban Land Use – Transportation Models as Tools for Evaluating Vehicle Travel Reduction Strategies', in U.S. Federal Highway Administration (ed.), *Land Use Compendium*, prepared for the Travel Model Improvement Program, Washington, D.C.

Stern, D. I. (2004), 'The rise and fall of the environmental Kuznets curve', *World Development*, **32**, pp. 1419–1439.

Stren, R. and Halfani, M. (2001), 'The cities of Sub-saharan Africa: from dependency to marginality', in Paddison, R. (ed.), *Handbook of Urban Studies*, London: Sage, 466–485.

Sun, X., Katsigris, E. and White, A. (2004), *Meeting China's Demand for Forest Products, An Overview of Import Trends, Ports of Entry, and Supplying Countries, with Emphasis on the Asia-Pacific Region*, Washington D.C.: Forest Trends, the Center for Chinese Agricultural Policy and the Center for International Forestry Research, 31.

Tabak, F. and Crichlow, M.A. (eds) (2000), *Informalization: Process and Structure*, Baltimore: Johns Hopkins University Press.

Takahashi, J. and Sugiura, N. (1996), 'The Japanese urban system and the growing centrality of Tokyo in the global economy', in Fu-Chen, L. and Yue-Man, Y. (eds), *Emerging world cities in Pacific Asia*, Tokyo, New York, Paris: UNU Press, pp. 101–143.

Tardanico, R. and Menjivar-Larín, R. (eds) (1997), *Global restructuring, employment and social inequality in urban Latin America*, Boulder, CO: Lynne Rienner.

Tarr, J. (1999), 'Decisions about wastewater technology, 1850–1932', in Roberts, G. R. (ed.), *The American Cities and Technology Readers*, London: Routledge and the Open University, pp. 154–162.

Tarr, J.A. (1978), 'Transportation Innovation and Changing Spatial Patterns in Pittsburgh, 1950–1934', in *Essays in Public Works History*, Chicago: Public Works Historical Society.

Tarr, J.A. (1996), *The Search for the Ultimate Sink, Urban Pollution in Historical Perspective,* Akron: University of Akron Press.

Taylor, P.J. (1995), 'World cities and territorial states: the rise and fall of their mutuality', in Knox, P.L. and Taylor, P.J. (eds), *World Cities in a World-System,* Cambridge: Cambridge University Press, pp. 48–62.

Telles, E. (1995), 'Structural Sources of Socioeconomic Segregation in Brazilian Metropolitan Areas', *American Journal of Sociology,* (100)5, 1199–1223.

The Economist (1998), 'Dirt poor: a survey of development and the environment', *The Economist,* March 19, 3–16.

The Journal of Urban Technology (1995), 'Special Issue: Information Technologies and Inner-City Communities', **3** (1-9), Fall.

The World Bank Group (2004), *Urbanization and cities: Facts and figures,* http://www.worldbank.org/urban/facts.html, retrieved on February 13, 2004.

Thomas, J.J. (1995), *Surviving in the city: the urban informal sector in Latin America,* London: Pluto Press.

Thomas, J.M. and Ritzdorf, M. (eds) (1997), 'Urban Planning and the African American Community: In the Shadows', *Thousand Oaks*: Sage.

Thompson, S. (2003), 'Planning and Multiculturalism: A Reflection on Australian Local Practice', *Planning Theory and Practice,* **4** (3).

Thompson, S., Dunn, K., Burnley, I., Murphy, P. and Hanna, B. (1998), *Multiculturalism and Local Governance: A national perspective,* Sydney: New South Wales Department of Local Government, Ethnic Affairs Commission of New South Wales and University of New South Wales.

Tian, X. (1998a), *Population Problem,* Beijing: Population Research Institute of PRC Academy of Social Sciences.

Tian, X. (1998b), 'Population Problem', Population Research Institute of PRC Academy of Social Science, *Urban Planning Review,* **3**.

Tocqueville, A. de (1988), *Democracy in America,* New York: Harper Perennial.

Tostensen, A., Tvedten, I. and Vaa, M. (2001), 'The urban crisis, governance and associational life', in Tostensen, A., Tvedten, I. and Vaa, M. (eds), *Associational Life in African Cities: Popular Responses to the Urban Crisis*, Uppsala: Nordiska Afrikainstitutet, pp. 7–26.

UN United Nations (1998), 'Fifty years of the economic survey', *Economic Survey of Latin America 1997–98,* 343–368.

UN United Nations (1999), *World Population Prospects: The 1999 Revision,* New York: UN Population Division, Department of Economic and Social Affairs.

UN United Nations (2001), *World Urbanization Prospects: The 1999 Revision,* New York: UN Department of Economic and Social Affairs, Population Division.

UN United Nations (2002), *World Urbanization Prospects: The 2001 Revision,* New York: United Nations, Department of Economic and Social Affairs.

UNCHS United Nations Centre for Human Settlements (2001), *Cities in a Globalizing World, Global Report on Human Settlement in 2001,* London: Earthscan.

UNCHS United Nations Centre for Human Settlements (2001), *State of the World Cities 2001,* Nairobi: UNCHS.

UNCTD United Nations Conference on Trade and Development (2001), *World Investment Report 2001,* New York: United Nations.

UNCTD United Nations Conference on Trade and Development (2002), *Trade and Development Report 2002,* New York: United Nations.

UNDIESA United Nations Department of International Economic and Social Affairs (1989), *Social situation 1989,* New York: United Nations.

UNDP (1994), *Human Development Report.* http://www.undp.org/hdro/hdrs/1994/english/94.htm, (retrieved on Sep 24, 2004).

UNDP United Nations Development Programme (2000), *Human Development Report 2000: Human Rights and Human Development,* New York: Oxford University Press.

UNECLAC United Nations Economic Commission for Latin America (1996), *Situación de la vivienda en América Latina y el Caribe,* Santiago.

UNECLAC United Nations Economic Commission for Latin America (2002), *Panorama Social de América Latina, 2001–2002,* Santiago.

UNECLAC United Nations Economic Commission for Latin America (2003), *Balance preliminar de las economías de América Latina y el Caribe 2002,* Santiago.

UNECLAC/UNCHS United Nations Economic Commission for Latin America/ United Nations Centre for Human Settlements (2000), *From rapid urbanization to the consolidation of human settlements in Latin America and the Caribbean: a territorial perspective,* Santiago.

UNESCAP United Nations Economic and Social Commission for Asia and the Pacific, *What is good governance?,* http://www.unescap.org/huset/gg/governance.htm, retrieved on February 20, 2004.

UN-HABITAT United Nations Human Settlement Programme (2002), *Global Urban Indicators Database: Version 2,* Nairobi: UN-HABITAT.

UN-HABITAT United Nations Human Settlement Programme (2003), *The Challenge of Slums: Global Report on Human Settlements 2003,* London: Earthscan.

UN-HABITAT United Nations Human Settlement Programme (2003), *Water and Sanitation in the World's Cities,* London: Earthscan.

United Nations Population Division (2001), *The state of World Population,* http://www.unfpa.org/swp/swpmain.htm, retrieved on February 12, 2004.

Van Paassen, C. (1962), Geografisch structurering en oecologisch complex, een bijdrage tot de geografische theorievorming, *Tijdschrift van het koninklijk Nederlands Aardrijkskundig genootschap (KNAG), 79,* 215–233.

Varley, A. (2002), 'Private to Public: Debating the Meaning of Tenure Legalization', *International Journal of Urban and Regional Research, 26* (3), 449–461.

Veenhoven, R. (2000), 'Well-being in the welfare state: level not higher, distribution not more equitable', *Journal of Comparative Policy Analysis, 2,* 91–125.

Vernon, R. (1966) 'International investment and international trade in the product cycle', *Quarterly Journal of Economics, 80,* 190–207.

von Weizsäcker, E.U., Lovins, A.B. and Lovins, L.H. (1997), *Factor Four: Doubling Wealth – Halving Resource Use,* The new report to the Club of Rome, London: Earthscan.

Wacquant, L.J.D. (1996), 'The Rise of Advanced Marginality: Note on its nature and implications', *Acta Sociologogica, 39* (20), 121–139.

Wakeman, Jr., F. (1993), 'The Civil Society and Public Sphere Debate: Western Reflections on Chinese Political Culture', *Modern China, 19* (2), April, 108–138.

Wallace, M. and Milroy, B. (1999), 'Intersecting claims: planning in Canada's cities', in Fenster, T. (ed.), *Gender, Planning, and Human Rights*, London: Routledge.

Wallerstein, I. (1990), 'Culture as the Ideological Battleground of the Modern World-System', in Featherstone, M. (ed.), *Global Culture: Nationalism, Globalization and Modernity*, London, Newbury Park, and Delhi: Sage.

Watson, V. (2002), *Change and Continuity in Spatial Planning: Metropolitan Planning in Cape Town under Political Transition*, London: Routledge.

WCED The World Commission on Environment and Development (1987), *Our common future*, Oxford: Oxford University Press.

Webster, D. (2001), 'Inside Out: Peri-urbanization in China', unpublished chapter, Stanford University Asia Pacific Research Center.

Weggel, O. (1999), *Alltag in China*, Hamburg: Hamburg Institut für Asienkunde.

West, C. (2001), 'Is it all coming together? Further thoughts on urban studies and the present crisis', *CITY*, **5** (3).

White, G. (1996), 'The Dynamics of Civil Society in Post-Mao China', in Hook, B. (ed.), *The Individual and the State in China*, Oxford: Clarendon Press, pp. 196–222.

White, G., Howell, J. and Xiaoyuan, S. (1996), *In Search of Civil Society: Market Reform and Social Change in Contemporary China*, Oxford: Clarendon Press.

WHO World Health Organization (1999), 'Health: creating healthy cities in the 21st Century', in Satterthwaite, D. (ed.), *The Earthscan Reader in Sustainable Cities*, London: Earthscan, 137–172.

Winarso, H. (1999), 'Private Residential Developers and the Spatial Structure of Jabotabek', in Chapman, G.P., Dutt, A.K. and Bradnock, R.W. (eds), *Urban Growth and Development in Asia, Volume I: Making the Cities,* Aldershot: Ashgate, pp. 277–304.

Winner, D. (2000), *Brilliant Orange: The Neurotic Genius of Dutch Football,* Bloomsbury.

World Bank (1995), *World Development Report 1995*, Oxford: Oxford University Press.

World Bank (2000), *World Development Report 2000/2001*, Oxford: Oxford University Press.

World Bank (2000a), *World Development Report 1999/2000: entering the 21st Century,* New York: Oxford University Press.

World Bank (2000b), *Cities in Transition: World Bank Urban and Local Government Strategy*, Washington, DC: World Bank.

World Bank (2002), *Globalization, growth, and poverty: building an inclusive world economy,* World Bank and Oxford: Oxford University Press.

Wright, A. and Wolford, W. (2003), *To Inherit the Earth: The Landless Movement and the Struggle for a New Brazil*, Oakland, CA: Food First Books.

WTO World Tourism Organization (1995), *Lanzarote Charter for Sustainable Tourism*, Madrid.

Xu, L.C. and Zou, H. (2000), 'Explaining the changes of income distribution in China', *China Economic Review,* **11**, 149–170.

Xu, W. and Tan, K.C. (2001), 'Impact of Reform and Economic Restructuring on Rural Systems in China: A Case Study of Yuhang, Zhejiang', *Journal of Rural Studies,* **12**, 165–181.

Yaeger, P. (1996), *The Geography of Identity*, Michigan: University of Michigan Press.

Yang, G. (2003), 'The Co-evolution of Internet and Civil Society in China', *Asian Survey*, **43** (3), May-June, 405–422.

Yiftachel, O. (1992), *Planning a Mixed Region in Israel: The Political Geography of Arab-Jewish Relations in the Galilee*, Aldershot: Avesbury.

Yiftachel, O. (1996), 'The internal frontier: territorial control and ethnic relations', *Regional Studies*, **30** (5), 493–508.

Yiftachel, O. (2004), *Ethnocracy: Land, Politics, and Identities in Israel/Palestine*, Philadelphia: University of Pennsylvania Press (forthcoming).

Yitna, A. (2002), 'Levels, trends and determinants of fertility in Addis Ababa, Ethiopia', *Journal of African Policy Studies,* **12** (2).

Yoshida, S. and Ma, L.-Z. (2000), 'China', in Policy Office of the Aging of Society, Director-Secretariat of the Management and Coordination Agency, Tokyo (ed), The Status of Aging in the Population and Measures for an Aged Society in Various Countries, pp. 63–74.

Yu-ping Chen, N. and Heligman, L. (1994), 'Growth of the world's megalopolises', in Fuchs, R.J., Brennan, E. *et al.* (eds), *Mega city growth and the future*, Tokyo, New York, Paris: UNU Press, pp. 17–31.

Zhang, L. (2001), *Strangers in the City: Reconfigurations of Space, Power, and Social Networks within China's Floating Population*, Stanford, CA.: Stanford University Press.

Zubrin, R. (with Wagner, R.) (1997), *The case for Mars,* New York: The Free Press.

Index